T0199359

The Economics of
Livestock Systems in
Developing Countries

The Economics of Livestock Systems in Developing Countries

Farm and Project Level Analysis

James R. Simpson

Routledge
Taylor & Francis Group

LONDON AND NEW YORK

First published 1988 by Westview Press, Inc.

Published 2019 by Routledge
52 Vanderbilt Avenue, New York, NY 10017
2 Park Square, Milton Park, Abingdon, Oxon OX14 4RN

Routledge is an imprint of the Taylor & Francis Group, an informa business

Copyright © 1988 Taylor & Francis

All rights reserved. No part of this book may be reprinted or reproduced or utilised in any form or by any electronic, mechanical, or other means, now known or hereafter invented, including photocopying and recording, or in any information storage or retrieval system, without permission in writing from the publishers.

Notice:
Product or corporate names may be trademarks or registered trademarks, and are used only for identification and explanation without intent to infringe.

Library of Congress Cataloging-in-Publication Data
Simpson, James R.
The economics of livestock systems in developing countries.
(Westview special studies in agriculture science and policy)
Bibliography: p.
Includes index.
1. Livestock—Economic aspects—Developing countries.
2. Livestock—Developing countries—Cost effectiveness.
3. Agricultural development projects—Developing countries—Cost effectiveness. 4. Agricultural systems—Economic aspects—Developing countries. I. Title.
II. Series.
SF55.D44S56 1988   338.1′36′0091724   87-25402
ISBN 0-8133-7510-X

ISBN 13: 978-0-367-29156-3 (hbk)
ISBN 13: 978-0-367-30702-8 (pbk)

TO

ITSUKO

CONTENTS

LIST OF TABLES

LIST OF FIGURES

LIST OF APPENDIXES

PREFACE

Improvement of the world's livestock industry necessitates input from many directions. Planning by a host of national and international agencies is required in order to provide optimal stimulus in very diverse areas ranging from formulation of laws and incentives that stimulate competition yet prevent undue competition, to the optimal provision of credit. Planning is needed to carry out a complex array of interrelated research and development projects, and it is necessary to be sure that investments and operational procedures are cost effective.

One fundamental aspect to improvement of any country's livestock industry is capital investment and optimal allocation of resources, whether they be through large-scale, government-initiated projects, such as rangeland improvement and irrigation, or very small-scale, producer-oriented and financed ones, such as cross-fencing of pastures, drinking water system improvement or rice straw treatment to improve feeding value. It would seem that investment and associated economic analyses within the livestock industry would be rather straightforward and simple, especially because several good farm management textbooks and a number of articles and books about improvement of the livestock industry in developing countries are now available. However, review of the literature indicates that no one reference exists that is specifically aimed at setting forth frameworks and methods for evaluation of investments and associated economic decisionmaking in the livestock industries of developing countries. That is the purpose of this book.

Economists will find this book useful because it synthesizes much material into a cohesive whole--material that is often presented in a cursory manner or overlooked as emphasis has shifted to sophisticated quantitative techniques. Even more serious is that students of economics seldom have training in animal husbandry, and thus a book that weds the two fields in a development perspective is needed. Perhaps an even greater audience is the animal scientist or administrator who wants a guide to the basic principles and methods of agricultural economics that apply to the livestock industry but is not interested in all the details, ramifications, and theoretical constructs behind investment decisionmaking. It is fair to say that a large group exists that would like a view of the overall picture and, within that, the why and how of the analytical procedures. These are the target groups for this book.

Evaluation of livestock-oriented investments and development programs requires an in-depth understanding of livestock systems themselves. As a result, one purpose of this book is to explain how livestock systems can be described for analytical purposes through an economics framework. Coupled with this is another objective--to set forth methods (approaches) and methodologies (philosophies) about farm-level analysis of livestock and resource-use problems in a systems context. Consequently, this book is designed to be a companion to livestock production textbooks written by animal scientists and to farm management textbooks by economists. The book is oriented to problems faced in developing countries, but much of the material is relevant to countries at all stages of economic development.

A third reason to write this book is to provide guidelines for micro-level analysis in livestock project development. Included here is the aim of expanding on the method known as farming systems research and extension (FSR/E) in livestock research programs. Recognition of FSR/E complements the farm management approach that has become a handmaiden of commercial farming. A major purpose in this regard is to explain how these traditional farm management methods and theory can be linked with FSR/E principles to provide a unified approach for developing-country livestock problem analyses. The idea is to show how the principles can be applied to any size or type operation, whether it be at the commercial or subsistence levels or anywhere in between.

Effort is given in this book to highlight the complementary roles often played in livestock systems such as meat and milk, traction and milk, or ownership of several types of animals. This is a handbook on farm-level livestock industry investment and resource analysis, yet this is not a "cookbook" because a major objective is to teach basic principles and thus application. The objective is to reach a wide and diverse audience working on developing-country livestock problems about how to conceptualize farm-level investment and related economic problems.

Credit goes to Toni Glover and Lavon Mikell for many revisions on the word processor. Thanks are due Westview Press staff for encouragement and skill in editing and guiding the manuscript through the editorial and production process.

Recognition is due to numerous people who assisted in preparation of this book: Allen Tillman, Peter Hildebrand, Robert Waugh, Guy Henry, Phylo Evengelou, John DeBoer, Patrick Moore and many graduate students. Most important appreciation and special thanks are due my wife Itsuko for her patience and continual encouragement.

James R. Simpson
Gainesville, Florida

CHAPTER 1

LIVESTOCK SYSTEMS IN INVESTMENT
AND ECONOMIC ANALYSIS

Considerable progress has been made in improving livestock production and marketing at the world level during the past several decades.
But the pattern of improvement both between countries and within
commodities is quite uneven. Some countries are highly automated with
great dependence on mechanical and electronic devices. All signs are
that the dependence of these countries on new technology and related
capital investments will continue at the same rate or even faster than in
the past. At the other extreme are whole countries, or regions within a
country, in which only the most basic-subsistence level technology and
management are employed.

Without doubt, the state of technical knowledge about livestock
production and the marketing of its products far exceeds human capability
to absorb it. As a result, the adoption gap in determination of
"appropriate technology" and integration of it is probably wider today
than ever in both the developed and developing world. One major reason
for this gap is lack of understanding about--and use of--basic investment
criteria, resource use principles and economic analysis at the livestock
production level by any of the interested groups: development specialists,
livestock producers, financial intermediaries and policymakers.

There are many approaches to presentation of materials about
investment analysis at the farm-level for the livestock industry in
developing countries. The one chosen for this book is to place economic
decisions within the context of livestock systems, problem definition and
human factors. For that reason, this first chapter is mainly devoted to
providing a framework to delineate livestock systems. The problems dealt
with are mainly confined to investments and resource allocation. Many
references are made in this chapter and this book to systems and
production practices in developed countries as one objective is to provide
information about technology which may or will have relevance to
developing countries. In effect, the emphasis in this book is on
development and change rather than description of what is currently
practiced.

Systems Analysis and Investment Decisionmaking

Investment decisions run the gamut from very simple ones that require virtually no quantitative analysis to those that can take large teams months to evaluate. An example of a relatively simple one is the decision by a single producer whether to feed pigs in a confinement house or to continue a backyard operation. At the more complex extreme is analysis of the extent to which a country should emphasize livestock product exports and, if so, which ones, from what region and how. In this case, analysis of entire production and marketing systems may be necessary.

The analysis of a large problem like export promotion may invoke various aspects of production and marketing. Ideally it would be possible to have physical analyses available of the various systems as well as associated cost and return information so that an investment could be conceptualized, appropriate data entered in a model (probably a mathematical one run on a computer) and the effect on resource use, outputs and cost provided. Unfortunately, even though the subdiscipline known as systems analysis is fairly well advanced, such models are just in the most elementary stages and are only available for quite restricted applications. Thus, livestock industry analysts are forced to fall back on logical and theoretical conceptualization of problems and, from there, on rather straightforward quantitative analysis of individual projects or investments. If the investments are to be carried out by private individuals, the analytical procedure is generally called capital budgeting analysis. The term *project analysis* (what was previously called cost/benefit analysis) is used for public investments.

It is commonplace that specialists from numerous fields be involved in project analysis. Each participant will have a different viewpoint about the "best" investment or the "best" way to organize any project. Furthermore, teachers will find that students, whether they be at the university level or producers themselves, will have many angles from which they view investment decisions. Because of this, the approach taken in this book is to provide concepts that can be applied across a wide variety of situations ranging from livestock systems that heavily use purchased input (commercial operations) to subsistence-level ones with virtually no purchased inputs.

The quantitative methods or tools for investment analysis are reasonably straightforward, although application to particular livestock problems is complicated by specialized factors inherent to the industry. But for most professionals the greater difficulty comes in problem definition and analysis of it within a wider production economics framework. As a consequence, highlights about the subdiscipline known as systems analysis as they relate to livestock are now presented. Then, attention turns to the scientific method for research, that is, the way in which problems can be conceptualized. These concepts are useful as background to application of investment analysis in an overall industry development context; they are crucial to conceptualize specific analyses as part of a fluid system in which today's answers are tomorrow's inputs.

The term *systems* in a research mode has come to be thought of as a holistic approach, as contrasted with a disaggregated procedure (McGrath, Nordlie and Vaughn, 1973). It is fundamentally a research tool, but has as much applicability in the everyday world of planning and project design as in the most basic university context. Systems research is a goal-oriented process of study through specification of the relevant system, performance of that system, and environmental variables. Included is determination of the existence, degree and form of relationship among the specified variables and use of the information to arrange or redesign the system so that it operates in the optimal (most efficient) way in terms of the specified objectives.

Dillon has written (1976) that the research process essentially involves four stages: development of a research model, synthesis and information interpretation, development of amended or redesigned models, and evaluation of the system's performance. This view, which is typical of systems methodology, moves scientific research from a reductionistic and mechanistic approach to an expansionistic one based on synthesis. It is an attempt to gain understanding of a system's structure or parts by understanding the way in which the whole system operates. Parts of a system are not judged on how well they operate, but rather in terms of their contribution to the whole (Wright et al., 1976; Johnson and Rausser, 1977).

The developments in animal science and forages, while generally described in terms of specific advances, also reveal concern with systems methodology (Koch and Algeo, 1983; Reid and Klopfenstein, 1983). Most systems analyses are at the biological level, such as modeling cow-herd performance, or at the industry level, such as comparison of various dairy-cow systems. But the concepts can also be applied to a farm firm. As an example, in construction of animal raising facilities it could be considered from the viewpoint of the impact on the entire farm enterprise, such as input use, cash flow, labor use and complementarity with other enterprises. Then the analysis is being carried out in a systems context.

The systems approach is continually evolving in response to outside influences as well as theoretical developments--that is, the approach is a building-block concept to the history of ideas. Thus, it happened that development of large computers led to interest in large-scale models. More recently, the special concerns about small farmers have led to the subdiscipline known as farming systems. Adherents of this latter group use diagrams rather than computer modeling at this stage to represent systems because of the holistic nature involved, much of which is difficult to quantify.

Overall, much more attention is being given to thinking along systems lines than ever before--especially in view of the astounding number of new technologies now available and ones that will be perfected and released in the next few years. The greatest contribution by animal-related researchers and action agents, especially in the developing countries, will be on-farm testing of the new technologies--and this implies a problem-oriented systems approach (Raun and Turk, 1983). In effect, as will be discussed in the last chapter, technology availability (and this includes even the most basic management techniques) is generally not the major constraint to expanded livestock-related production or to improvement of nutritional diets in developing countries. Rather, the major constraint is proper problem definition, evaluation of the technologies that are appropriate, identification of the optimal input mix and then development of appropriate means to adapt them (Simpson, 1983). In this sense, improvement of the livestock industry is basically a human oriented problem, not a technical one.

Research Methods in Livestock Analysis

The way of thinking in science is called its methodology or its research method. Many philosophers and scientists have written about the research method, but one method that seems to have much relevance to livestock related investment and related economic analysis in a systems context was set forth by Kemeny, who divided the analyst's universe into the world of facts and the world of experimentation (Figure 1.1). The first step is to gather basic information, which may come from interviews with livestock owners or others, a literature review, statistical sources, and so on. This information can be considered the "facts."

Once the basic facts are assembled, inductive reasoning is used to develop theories about the problem the researcher has first identified. In practice, the theories take the form of models that are then used, through the method of deduction, to predict the way in which the real world could be expected to operate under given conditions. The predictions or modeling results are then verified. At this stage the researcher is led back to the world of facts, from which the inductive process is used once again to amend and redesign the model. The methodological key for successful systems work is to recognize that research--whether it be very theoretical or quite applied and action oriented--is an ongoing process of continually redefining the system in light of new facts and redefined objectives.

The scientific method is an integral part of systems science and involves continual moving back and forth between the world of facts and the world of theory. Fitzhugh (1975) points out the relation between modeling and the scientific method in systems science as:

Develop hypothesis <==> Develop model
Test hypothesis <==> Validate model
Revise hypothesis <==> Revise model
Test revised hypothesis <==> Validate revised model

In effect, the model is the hypothesis to be tested, and, as such, the systems approach is more than a collection of mathematical techniques; the systems approach is a philosophy of problem identification, evaluation and solution.

A somewhat confusing term is the term model, which is commonly defined as a synthetic representation of a system. To many, a model is a set of equations that abstract from reality; to others a model must be a major effort describing all of the various ramifications from alternative courses of action. Simulation is an example of one modeling technique that, in the strict sense, is an attempt to approximate real conditions as closely as possible. Another example is linear programming, which is an optimization technique. These tools are powerful but often require large amounts of research time and funds.

Budgeting can often provide, as will be shown in later chapters, just as meaningful and useful answers for development projects. Thus, because budgeting is often a far less costly technique, it is usually the most appropriate one. The problem is to determine when to use each one. Also, the same data are frequently needed for both budgeting and larger modeling efforts. But are budgets models?

The question of whether budgets are models can lead to considerable discussion with strong arguments on both sides. A rather pragmatic approach is taken in this book--budgets are models if the intent is to use them as a "synthetic representation of a system" (a system is "a collection of interrelated components or elements with a common purpose"). In other words, although budgeting as used by an accountant is certainly not a modeling effort, in the hands of a livestock systems analyst budgeting can-be just as potent a tool as a complex set of mathematical equations to describe, analyze, and solve livestock systems problems.

Let us return to the discussion of the scientific method for a concrete example about how budgets can be used as models in farm-level decisionmaking. Suppose that a small farmer in a developing country is being helped to determine if a 10-sow total confinement feeding operation based on sale of weaned feeder pigs should be developed. The new enterprise will replace the 2-sow, low-input operation currently employed in which the animals basically run free and are only cared for in the most subsistence manner. The first step following the outline of Fitzhugh is to develop a hypothesis, in this case that the return to labor (or any other measure) will be greater than a certain predetermined amount. That leads to developing a model that in this case is deemed to be a cost-and-returns budget and a cash-flow analysis.

The model (budget) is built and validated by reviewing it with friends, researchers, government development agents, and others. The analysis of how well the pig operation fits in the farm's system (a holistic perspective) may lead to revision of hypotheses about the benefits from the operation, in which case the model may have to be revised. For example, it may be that the total net income from 10 sows is insufficient to justify the time expended. But, a 20-sow farrow to finish (that is, all phases to slaughter weight) permits specialization, which justifies additional training and a bank loan. Costs are determined to be substantially lower due to discount prices for volume feed delivery, and total revenue per animal is estimated to increase substantially by sale of a regular supply directly to a butcher rather than to itinerant traders. The revised hypothesis leads to a revamped model (budget) and, if necessary, to further modifications until the farmer either accepts or rejects the proposed new system.

The method outlined in Figure 1.1 can also be used as a means to understand the example problem's analytical procedure. In this approach, the first step is to develop the idea and conceptualize a plan. That is the process of gathering information in the world of facts. From there, the process of induction leads to construction of a budget (that is, a model that corresponds to theories that a more abstract analyst would develop in the world of experimentation and analysis). The method of deduction is used to make predictions about how the farm family will be affected if the 10-sow/feeder-pig operation is adopted.

The predictions are verified by returning to the world of facts through interviews with friends and others and by considering quantifiable as well as nonquantifiable factors impinging on the system. From this additional information a new budget, as described earlier, is constructed.

The discussion about analytical methodologies may seem far removed from practical day-to-day work. But in reality, this is the basic foundation for an orderly conceptualization of problem definition and analysis. Smooth execution of the principles will lead to efficient analysis of all types of livestock problems--even those never before encountered. Now, prior to delving into the intricacies of using budgeting as a modeling tool, we need to lay the groundwork for livestock systems design efforts by discussing factors that influence their evaluation.

Factors Influencing Livestock Systems

Livestock systems evolve due to a host of factors. The systems are not static, but change over time in response to technological, political and economic forces. This section's purpose is to examine the factors that act as stimulants or constraints because understanding these factors will help explain why particular types of livestock production systems have evolved as they have.

Land characteristics act as a prime determinant of livestock production systems because highly fertile land will more likely be devoted to agriculture while poorer quality land will be used for livestock production. Examples of this comparative advantage concept are the deserts of Mexico or northern Africa, which have little suitability for farming. As a result, raising livestock is generally the most justifiable economic alternative. But, cow/calf operations are found on quite good farming land as well. Why?

Production complementarity occurs in many farming areas. For example, in the midwestern United States corn farmers have the option of selling their harvest or marketing it through cattle (Boykin, Gilliam and Gustafson, 1980). Traditionally, many of these entrepreneurs have chosen to feed their grain to cattle as it has provided a source of employment during the slack winter months as well as a means to market corn. More recently, some farmer-feeders have joined together to fatten cattle on a cooperative basis, thus taking advantage of size economies. Other farmers frequently maintain a small herd of beef cattle to provide calves for feeding operations. In Europe, dual purpose (milk and meat) cattle are excellent examples of complementarity between milk, beef and crops. These farmers also frequently feed-out (fatten) progeny as a means to market farm products and by-products. Dual-purpose cattle systems are widely used in Central America as a means to reduce risk from adverse price movements of either commodity and as a means to generate daily cash flow (CATIE, 1983; Pearson de Vaccaro, 1982).

Tradition and social conditions such as population density and land resource distribution also strongly affect land use. Peasants in central Mexico farm very poor, rocky, rough semidesert land that apparently is best suited for raising cattle, goats or sheep. It is true that on a national or international scale the land clearly should be in livestock. But people cannot be moved around like pawns on a chessboard. Thus, if farmers have little alternative use for their family's labor, a higher total net return can be realized from farming rather than cattle raising. The return on a per-hour-of-time-invested basis is likely lower from farming than cattle raising, but the total return is higher. One of the objectives in livestock development projects is to determine the highest and best use of land on a national scale subject to individual cost/benefit ratios and then to design the projects accordingly.

The level of a nation's economic development, combined with cultural mores, has a strong influence on livestock systems. Much of Africa's land is inhabited by tribal nomads or seminomads who have little effective interest in expanding production (Simpson and Sullivan, 1984). Effective interest means that interviews with the cattle owners may indicate an apparent desire to improve production, but that the desire is not strong enough to be translated into action. In many potential exporting countries, social constraints prevent maintenance of packing plants at international standards, thus excluding these countries from competing in world beef markets.

Location is another factor influencing livestock raising systems. There are various areas, such as Sweden, in which livestock production must be subsidized for operations to compete with exporting countries such as Australia or Uruguay. Futhermore, the law of comparative advantage strongly dictates the type of commodity that can be produced in a certain area. For example, Japan has a high effective demand for beef, but a low per capita consumption of it (Simpson, Yoshida, Miyazaki and Kada, 1985). Instead, the Japanese have concentrated on rice because this is the highest and best use of their very limited natural resources. In contrast, the Swedish recognize that cattle raising is a better alternative in some regions than are other options, such as timber.

Another aspect of location is the prevalence of disease. Much of Central Africa is well suited to cattle raising, but the tsetse fly precludes cattle raising in many zones except under special conditions. Furthermore, prevalence of certain diseases such as rinderpest and foot-and-mouth disease prevents most African countries from fully competing in world trade.

Level of effective demand combined with national borders of nation states can effectively limit production increases in many countries. Per capita meat consumption in all of Africa and much of the Near East is very low because most individuals do not have the necessary income to purchase greater per capita quantities of meat than are presently taken. As another example, meat production may be essentially constrained to certain limits because there is no effective market for the product. This is the situation in the South American countries of Argentina, Paraguay and Uruguay. These countries, as well as most Latin American countries, could greatly expand beef production if a market and attractive prices prevailed.

Price of meat is an extremely important determinant of livestock raising systems. If meat prices are low, livestock owners cannot afford to invest as much in inputs as in the case of high prices. The production systems in Europe and the United States are quite capital intensive because prices are relatively high. In South America, high interest rates, inflation, lack of capital, low meat prices and low wage rates make the operations much more labor intensive (Valdez and Nores, 1979). Low prices also lead to a tendency for livestock raising to be essentially an extractive business because the return to management is low compared with alternative investment opportunities.

Government policies, as already mentioned, are one of the most important factors in the evolution of livestock-raising systems because the industry requires incentives to grow and improve. Most countries interfere with the free market to influence livestock prices in some manner (Simpson, 1984c). Frequently, retail or wholesale prices of some or all cuts are fixed. Some countries that export beef and cattle, such as Costa Rica, have a two price system, one for national consumption and the other for export. Others, such as those in the European Economic Community (EEC), provide subsidies to producers and exporters, thus disrupting trade and development that would otherwise occur in developing countries (Simpson and Hillman, 1985). An example already cited is the inordinate disadvantage faced by many developing countries in their attempts to develop domestic milk industries due to subsidies on milk powder exports by many countries, such as those from the EEC.

Another aspect of government policy relates to stability and provision of a favorable investment climate. Livestock production, especially cattle, requires substantial investment, and, due to the industry's biological nature, returns are often deferred for several years. If there is fear of expropriation, adverse government involvement, or radical shifts in policy, landowners will prefer to maintain most of their capital in livestock (relatively liquid assets) rather than in land improvements and other investments. Chile, Peru, Uruguay and Zimbabwe are good cases of countries in which government policies or political turmoil have periodically prevented the cattle industries from intensifying and evolving into more efficient forms.

Approaches to Systems Categorization

The way in which a livestock and agricultural system is categorized largely depends on the researcher's interest and on the problem at hand. For example, Grigg (1974) takes an evolutionary approach. Ruthenberg's (1971) interest is tropical systems. Kolors and Bell (1975) combine geography with environment and people. Tillman (1981) places livestock systems within a livestock industry development perspective. McDowell and Hildebrand (1980) focus on small-farmer integrated-crop-and-animal systems. Hoveland et al. (1977) approach the topic from a forage systems viewpoint. Bredahl, Burst and Warnken (1985) use cattle production systems as a basis to analyze the growth and structure of a country's livestock industry. Conrad (1984) divides systems into confined, semiconfined and nonconfined. Winrock International researchers have developed an aggregated classification that divides world livestock systems into three types: animal based, crop and animal based, and crop based. These researchers have identified nine subclasses in developing countries (Winrock, 1981; Fitzhugh, 1983). Review of these and other studies suggests that systems classifications are developed to meet specific needs.

A natural inclination or desire when working at the world level is to classify livestock systems by country. For example, it is possible to say that the U.S. beef cattle system is characterized by cow/calf operations that sell weaned calves to stocker operators after which the calves are finished in feedlots (Petritz, Erickson and Armstrong, 1982) or that Kenya has a substantial portion of its cattle kept for milk by pastoral peoples. Although this approach is useful as an introduction to the world's livestock business (Simpson and Farris, 1982), it is a rather cumbersome approach for analytical work due to the wide range of subsystems found in virtually every country in the world. Furthermore, livestock are more problem than site specific (as is the case in crops). Thus, for operational or analytical purposes, a more useful means to describe systems is according to activity, institutional arrangement, size and feed source.

Extensive Versus Intensive Cattle Systems

One useful way to broadly separate beef cattle systems is into extensive and intensive ones (Table 1.1). There is no solid definition of the difference, but it can generally be said that extensive operations are

Table 1.1. Factors for categorizing cattle systems

Extensive Operations	Intensive Operations

Activity

Cow/calf	Cow/calf
Integrated, cow/calf and growing	Integrated, cow/calf and growing
	Dairy
	Integrated farming and live-stock
	Integrated cattle and other livestock
	Backgrounding
	Stocker
	Finishing
	Forage
	Feedlots
	Dual-purpose cattle

Institutional Arrangement

Pastoral, communal grazing	
Family operations or sole proprietorship	Family operations or sole proprietorship
Corporate ranches	Corporations
Cooperatives	Cooperatives
State farms	State farms
Traditional versus modern	

Size

Very small (19AU or less)	Very small (19 AU or less)
Small (20-149 AU)	Small (20-149 AU)
Medium (150-399 AU)	Medium (150-399 AU)
Large (400-799 AU)	Large (400-799 AU)
Very large (800 AU or more)	Very large (800 AU or more)

Feeding

Native range	
Improved pasture	Improved pasture
Preconditioning	Preconditioning
Backgrounding	Backgrounding
Feedlot	Feedlot

AU = animal unit

ones that require a large geographic area per animal, while intensive operations refer to those with a high cattle/land density ratio. To avoid terminological confusion, the definition in this book will not refer to management levels or, as some researchers have, classify intensive operations as ones where feedstuffs are brought to the livestock (Preston and Willis, 1974). Rather, the definition will just mean a "high" cattle/land density ratio for given climatic conditions.

Extensive operations are usually found in areas that are only marginally useful for other agricultural operations. Dry zones such as deserts or semideserts, mountains, or rocky regions are usually utilized on an extensive basis. Arizona, a state in the arid southwestern United States, is characterized by extensive cattle operations due to low rainfall. South Florida, on the other hand, has an abundant rainfall but is also characterized by extensive ranching operations. This is partly due to poor soil, but is more a function of tradition, location in the extreme southeastern part of the United States, lack of a large cattle-feeding (finishing to slaughter weight) industry, and low alternative uses for the land. In addition an adverse fertilizer/beef price relationship has prevented much more intensive use of pasture land.

The dynamic nature of systems is also exemplified by Florida. Up until the early 1950s, there were few improved beef cattle and virtually all cattle were grazed on open, (unfenced) ranges. The 1950s, 1960s and early 1970s witnessed the adoption of intensive practices such as the planting of improved forages, advances in management practices, better nutrition, and upgrading through breeding (Simpson and Bordelon, 1984). In 1950, there were 3.7 hectares (ha) of agricultural land required per head statewide of cattle inventory. By 1982, just 1.4 ha were required, or only about one-third as much. Thus, as technological advances were made, population grew, and the transportation infrastructure developed, many of the cattle operations became much more intensive.

Extensive systems are characterized by cow/calf operations in which the primary activity centers around a breeding herd of cows kept for the production and sale of weaned calves. This type of system is found throughout the western United States, the dryer sections of Australia, and the more remote areas of Latin America. Distance from market and cattle price/transportation cost relationships are also important (Simpson and Steglin, 1981). Much of Latin America, Australia, and Africa has extensive operations that are integrated all the way from cow/calf through finishing--and all on forages. As will be explained in the chapter on cattle feeding, this practice prevails when cost of grain or other feedstuffs is high relative to cattle price.

Another factor to consider in categorizing cattle operations is institutional arrangements (Table 1.1). In most countries where free enterprise prevails, extensive operations are family owned partnerships, sole proprietorships, or corporate holdings. In contrast, in much of Africa and on communes in the rangelands of China and the USSR, a pastoral system dominates in which cattle are owned privately but herded on communally owned land. Generally, the land is held in trust by the federal government. The institutional arrangement in centrally planned economies may be a state farm in which most assets belong to the state or perhaps a cooperative with land held in trust by the state but all other assets owned by the cooperative. However, even the most doctrinaire communist countries, such as the USSR, allow limited private ownership of cattle and other livestock (Gray, 1981).

Size is the third major consideration in characterizing cattle systems. Although size classifications are necessarily arbitrary, the ones that have come out of the comprehensive cost-and-returns studies in the western United States (Boykin, 1968; Boykin and Forrest, 1971; Goodsell, 1974; Gray, 1968) can be taken as a reference point for areas that have extensive cow/calf operations such as much of Latin America, Africa, Oceania, and China. Small ranchers are those with 20-149 animal units (AUs), medium have 150-399 AU, large have 400-799 AU, while very large run 800 AU or more. In addition, a classification named very small can be added for enterprises with 19 AU or less. Researchers can devise their own classifications depending on research objectives and the area under study. It would be meaningless, for example, to hold to the foregoing classifications if virtually all owners in a study area had 19 AU or less.

Measurements with Animal Units

The wide differences in herd composition on any one operation due to the diversity of land types, classes of livestock, climatic conditions, land values, input costs and cattle prices have led researchers to rely on a standardized measure called animal units (AU).

The legal definition of an animal unit in the United States is a beef animal of more than six months of age, five mature sheep or goats, or one horse (Gray, 1968, p. 122). This is also the definition used by public land agencies in the United States when they determine grazing permits.

The total number of animal units is calculated by multiplying the numbers of each type animal by the appropriate coefficients. Although the coefficients are determined by the amount of forage an animal eats, which in turn is related to body surface, (cattle size) the following coefficients generally hold for Brahman or English breed cattle and small stock:

Type of Animal	Animal Unit Coefficient
Mature cow	1.00
Long yearling (one nearing 24 months of age)	0.80
Weaned calf	0.50
Unweaned calf	0.40
Pregnant heifer	1.00
Cow with calf	1.40
Mature bull	1.25
Two-year-old steers	1.00
Horse	1.25
Sheep	0.20
Goat	0.20

A procedure for converting cattle other than the ones given here to AUs, or for cattle whose size varies to any significant degree from English or Brahman cattle, is to use a formula that converts body weight to animal units. The formula is

$$AU = 0.01693 \ (W^{\frac{2}{3}})$$

$$= 0.01693 \ (\sqrt[3]{w})^2$$

As an example, assume that the average mature cow weight in a region being worked on is 343 kilograms (kg). The resulting AU would be

$$AU = 0.01693 \ (\sqrt[3]{343})^2$$

$$= 0.01693 \ (7)^2$$

$$= 0.01693 \ (49)$$

$$= 0.83$$

which is interpreted to mean that the AU equivalent is 0.83 rather than 1.0. All other animals would have to be adjusted as well to standardize the units. Special care would have to be taken with small stock as they will not necessarily be proportionally smaller.

The formula has the drawback of resulting in animal unit values that are somewhat higher than would be warranted by weight alone. Care must also be taken when using animal units to specify some nutritional level because practices vary considerably. The implicit assumption about the animal unit coefficients given previously is that they hold for "good" levels of nutrition--that is ones that when combined with other "good" management practices will result in an average (year after year) calf crop of more than 80 or 85 percent, where calf crop means the percentage of calves weaned from females in the breeding herd. Caution must also be exercised in defining the terms calf crop, mature animal, heifer, and so on, just as care must be taken in defining animal units. These caveats should not be considered burdensome because the essential point is usually to develop relationships between animals and different areas rather than precise totals.

Livestock analysts carrying out in-depth studies often break the year into months, with inventory on an animal unit month (AUM) basis, because livestock inventory numbers change throughout the year in response to the impact of seasons on forage availability, calving dates, livestock sales, transfer of stockers or breeding stock, and so forth. The use of AUMs is also especially useful in making comparisons between radically diverse areas. An example is a stocker operation in a desert or a mountainous region where cattle are only grazed for a few months a year contrasted with a tropical zone where one irrigated piece of land planted to a high yielding forage will permit continual grazing throughout the year.

Intensive Cattle Systems

Areas that lend themselves to intensive operations potentially increase the type and scope of livestock raising activities. For instance, cow/calf operations, just like the growing or finishing of steers, can be intensive, and the nature of that enterprise can change over time if product/input price ratios vary sufficiently. An example is reaction to the sharp rise of beef prices in the early 1970s when some cattlemen in the United States began keeping breeding cows in confinement and feeding them on feedstuffs harvested mechanically. More recently, the use of embryo transfers from high quality donor cows to inexpensive cows that then carry the embryo to birth is another means to intensify the cow/calf stage of beef production. It could also be argued that grazing beef cows on improved pastures is a means to intensify beef production. Another intensive type system frequently practiced in very cold areas is main-tenance of beef cows in confinement facilities during the winter.

Intensive integrated cow/calf and growing systems are common in Europe, much of Asia and to some extent in virtually every country of the world. The European cattle industry, for example, is almost ex-clusively based on dairy production (Poitevin, Mallard and Picon, 1977) because those operations tend to be small units (apart from the coopera-tive farms of East Europe) and are subjected to harsh winters, especially in the northern regions. About three-quarters of European cattle are dual purpose dairy/beef breeds because single-purpose beef-type cattle have not provided as high a total net farm income. In fact, apart from

England, France, Ireland and Italy, beef breeds have only made up 2-5 percent of all cattle in Europe.

Farming operations are most generally associated with more intensive beef cattle operations such as growing weaned steers or finishing them to slaughter weights. The dramatic expansion in maize production, especially for silage, was one of the reasons the EEC shifted from being the principal beef importing bloc during the 1960s to the world's second largest net beef exporter (after Australia) in the mid-1980s. Feedstuffs development is one reason feedlots have developed so rapidly in Europe. Such operations, whether conducted on pasture or in confinement facilities, have traditionally been a means to utilize slack labor, land or capital rather than being the focal point of the production unit.

In Europe, the feeding of dairy calves to slaughter weight either as vealers or full-grown animals is an example of the association that develops between farming and cattle raising. Another example is the common practice of finishing purchased steers on corn stored by farmers in the pampas of Argentina. Many farm operators in the southeastern United States background steers in the winter months on otherwise idle farm land planted to rye or ryegrass. In other areas of the world where draft power is still used, such as much of Africa, Asia and the Middle East, farmers keep triple-purpose cows for milk, draft and as sources of calves. The female calves are used as replacements while the males are castrated and used for draft purposes.

Other Livestock Systems

The foregoing discussion on beef cattle is sufficient to provide a sense of the way in which livestock systems can be categorized and analyzed. An expanded discussion on goat, sheep, buffalo, dairy cattle and swine systems will be presented later. Those topics are delayed to permit presentation of quantitative techniques so that examples of analysis can be presented at the same time.

CHAPTER 2

BUDGETING AND PRODUCTION
ECONOMIC THEORY

Livestock systems research, as described in this book, is composed of a descriptive and an analytical component. The major technique for analyzing a livestock system, whether it be for comparative purposes at the world level or for profitability improvement on a one hectare farm with just one dairy cow and two sows, is enterprise budgeting. If only one small aspect of the farm is being evaluated, then partial budgeting may be an appropriate approach. The guiding determinant for management strategies is production economics. This latter theoretical framework, plus enterprise budgets and cost curves, is explained and discussed in this chapter as it relates to livestock systems research. Application is provided in the next several chapters.

The Budgeting Process

Budgeting is a powerful economic tool for analytical work at the regional and national level as well as farm level (Figure 2.1). For example, budgeting is the proper approach to determine commission charges in a livestock marketing cooperative or to estimate cost per unit in an abattoir. Furthermore, budgeting is the foundation for much project and regional analysis, as will be shown in later chapters. Many sophisticated modeling and optimization techniques, such as linear, nonlinear, quadratic and dynamic linear programming, are based on the data generated in the budgeting process. It is a cornerstone for livestock system design improvements because success or failure of any innovation ultimately depends on costs and returns. Thus, the salient point is that livestock research analysts, whether they be economists or not, understand the basic principles of budgeting, its relation to economic theory, and when and how to apply it. We begin by distinguishing between accounting and economics.

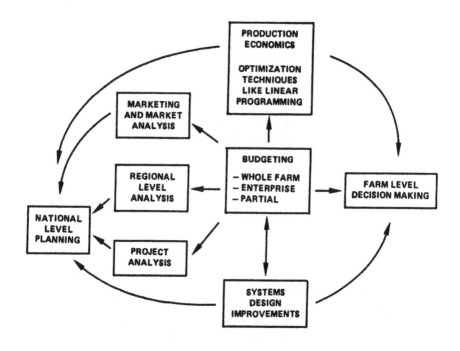

FIGURE 2.1   THE ROLE OF BUDGETING IN LIVESTOCK INDUSTRY
RESEARCH.

Accounting Records Versus Economic Budgets

Budgets can be developed in any number of ways depending on their purpose. Accountants, for example, are concerned with cash movements, dividing these into debits and credits. For accountants, the focus is basically on income and expenses in a particular time period. Accountants record these movements for control purposes, such as tax reporting or audit of a firm for inventory. But data from the records accountants keep can be used for analytical purposes in many situations (Osburn and Schneeberger, 1983). A somewhat more sophisticated step in accounting is development of an income statement, sources and uses of funds statement (SOF), net worth statement, and financial ratios (Penson, Klinefelter and Lins, 1982). This approach to budgeting is an important tool for solving certain problems in livestock systems research (Kay, 1981).

Economists, unlike accountants, are concerned with analysis (Richardson, Camp and McVay, 1982). In livestock-systems research economists may be called upon to analyze the economic health of a particular operation, in which case they might use traditional business management techniques, which begin with income, SOF and net worth statements (Harsh, Connor and Schwab, 1981). Then, economists probably would be interested in the economic potential of the operation under current management practices and under some alternative approaches. The problem at this point is to develop analyses that represent the operation in a typical mode to remove the influence of unusual practices in a particular year. This can be done by creating--that is, synthesizing--a budget rather than using the accounts per se.

One advantage of budgeting is that it then permits analysts to calculate costs and returns from readily accessible data rather than depending on the goodwill of livestock owners in disclosing their financial status. Furthermore, producers seldom have the data in a readily usable form. Perhaps most important is the lack of standardization among producers about which costs to include; thus, answers about net return or profit will vary greatly. But producers will usually readily provide the quantity of physical units and appropriate prices from which standardized budgets can be created. Another advantage of working through the budgeting process is it results in greater in-depth understanding of an operation.

Enterprise Budgets

Enterprise budgets--that is, cost-and-returns studies for one individual operation on a production unit--can be used separately or combined to provide a complete picture of the entire farm or ranch. Because farm-level systems research is related in general to the entire farm or operation controlled by an individual, many analysts will want to disaggregate the farm or system into each of the various enterprises and then combine them again into a whole. This is the approach in livestock systems research. If, of course, a production unit only has one operation, say beef cows, then that enterprise budget constitutes the farm budget. The more interaction among the various enterprises, the more reason to reaggregate them into an integrated whole.

This book is aimed at livestock systems, and, consequently, the examples in this chapter are primarily oriented toward livestock. The approach taken in these first few chapters is to develop budgets for enterprises that are reasonably sophisticated in the sense of using a wide variety of purchased inputs and then later showing how the concepts are applied to more subsistence-level situations. But it should be remembered that the concepts and techniques hold for all types of enterprises and can be applied, with some modification, to virtually any kind of business from a meat canning plant to production and marketing of goat's milk. The concepts also hold for any size operation and in any economy regardless of its primitiveness or sophistication. In effect, the objective is to first develop a clear understanding of theory and concepts and then to apply them to a wide variety of situations.

A budget is a systematic listing of income and expenses for a specified production period and is used for any or all enterprises in the operation: business financing, planning, management control and strategy development. Projected income is derived when expected yield is multiplied by expected price. Projected expenses are derived from variable and fixed costs. Net returns are calculated by subtracting expenses from gross income.

Budgets are developed from a variety of data sources. The most appropriate one depends on the researcher's purpose. In many cases a representative sample is needed to obtain an understanding of the study areas. This aspect, which could take place in the descriptive phase of the research, would, for example, probably be conducted via a survey. Another use might be management control or evaluation of one particular operation, in which case the operator's records would be the appropriate information source. A third use is to determine the potential for improved or alternative production practices in selected areas. Here, information probably would come from research stations, experience in other areas, or some of the better entrepreneurs in the study region.

Variable, Fixed and Production Costs

Considerable confusion exists about the terms variable costs and fixed costs, primarily because the categories can change between farms and even within a farm. The difficulty is in the rather nebulous definitions. For example, variable costs are those that vary within a production period; fixed costs do not vary within a production period. Seed is a variable cost before it is sown but fixed after being planted. Fertilizer becomes fixed after it has been spread. Thus, as the production period advances, more and more inputs become fixed.

Farm management economists distinguish between variable and fixed costs in order to use the amount of fixed costs in analysis of management strategy. For example, a very high percentage of a desert region cow/calf operator's costs are fixed because few production options are available. In contrast, a farmer in a subtropical area who occasionally grows postweaned calves is much more flexible and has fewer fixed costs than does the desert operator. Thus, type of resources is a factor in determining whether a factor is fixed or variable.

Another use of the term fixed relates to the way in which a fixed input is used. For example, average fixed cost (total cost divided by use) can be reduced by employing a large piece of equipment that is relatively efficient rather than several small implements. Another way to reduce average production cost is to use a fixed resource more hours to spread depreciation cost. Because time is so important, fixed and variable resources are also used to classify the length of the production period. These periods are the

Very short run	A time period so short that all resources are fixed
Short run	A time period in which at least one resource is varied while other resources are fixed
Long run	A time period in which all resources can be varied

Cash and Noncash Costs

Another confusing set of terminology is the difference between cash and noncash costs. Some inputs are easily separable while others are very nebulous. For example, calves purchased for fattening are clearly a cash cost, while there is no doubt that depreciation is a noncash cost. But what of management by the production unit's owner? That person spends time that could be used in other activities, such as working off the farm. Some wage should be received for time spent on his or her own operation, but unless the operator is rather sophisticated, usually no provision is made for a monthly wage to be recorded as a cash expense. Rather, what typically happens is the individual simply uses savings or part of production credit for personal expenses or takes part of the income when farm-produced commodities are sold. In this sense, because there is no direct, specific outlay, the cost moves from being a cash to a noncash cost. This terminological difficulty is overcome in this book by presenting the budgets with several other subdivisions and avoiding the terms cash and noncash costs.

Inputted Prices

Another difficulty encountered in budgeting livestock enterprises is placing a value on inputs and outputs that do not pass through market channels. Examples are seed that is raised on the farm and used for producing forage; calves weaned from the farm's cow herd and fattened to slaughter weight rather than being sold at weaning; or home consumption of meat from farm-raised animals. This is handled by inputting a current market price to each item in question.

Production Functions

The purpose of this section is to bridge the gap between the general discussion on budgeting and decisionmaking just presented and the use of production economics through some case examples given in the next chapter. Now, two concepts are discussed: production functions and cost curves. Both are effective tools to evaluate many livestock systems problems, primarily because these concepts provide a theoretical means to understand the production unit and the direction of impact from proposed changes. Thus, even though production and cost functions are seldom empirically measured, they are essential for the conceptualization process. Cost functions and production functions are derived from the same technical input/output relationship (Doll and Orazem, 1978). As such, these functions need to be studied together in order to understand and effectively utilize these important tools in livestock-systems analysis.

The production function describes the <u>rate</u> at which inputs are transformed into outputs in a given time period--for example, fertilizer into forage or grain to weight gain on steers. Some production functions of interest to cattlemen for evaluating production alternatives include relationships of gain by weight classes, sexes and breeds of cattle on different type forages. Other useful information would be alternative forage yield responses with different input combinations such as fertilizer, mowing or water. Additionally, data on interrelationships such as average daily gain from supplemental feeding and various carrying capacities can be useful. A typical production function is shown in Figure 2.2.

Symbolically, a production function can be written as

$$Y = f(X_1, X_2, X_3...X_n)$$

where Y is output and $X_1...X_n$ are different inputs (also called resources or factors of production). The "f" means "function of," showing that a unique amount of output is obtained from a given set of inputs. Some of the inputs can be varied while others are not allowed to vary. A vertical line separates the inputs into the ones being varied and the fixed ones. For example,

$$Y = f(X_1 \mid X_2, X_3...X_n)$$

means that X_1 is being varied while the others are being held constant.

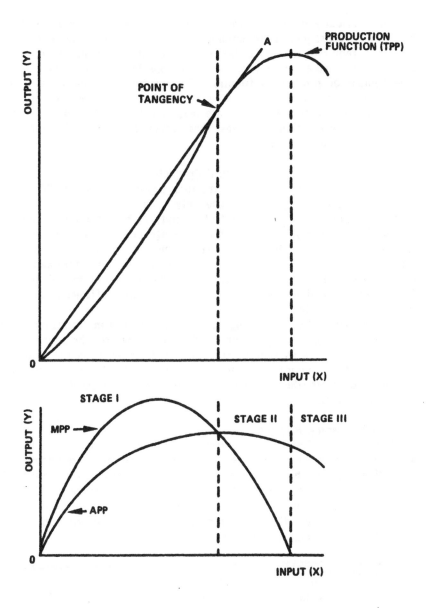

FIGURE 2.2 THE CLASSICAL PRODUCTION FUNCTION AND THE THREE PRODUCTION STAGES.

One difficulty in systems work is the paucity of production function data. As a consequence, researchers (not to mention livestock owners) are often required to utilize the few observations available for their estimates and make rather gross interpolations between the data points. Development of production functions for existing livestock systems and for possible ones is a major area of work for systems researchers. Because the production functions are mainly used by economists but are derived from production situations, these functions are best developed as an interdisciplinary team effort.

Total, Average and Marginal Physical Products

The output, Y, is often called the total physical product (TPP). The average physical product (APP), also only a physical measure, is calculated by dividing the total amount of output, Y, by the total amount of variable input. The APP measures the average rate at which an input is transformed into a product and is an efficiency measure because efficiency is measured by dividing output by input. Thus, as more and more input is used, efficiency eventually declines so the rate of growth in TPP also begins to decline (Figure 2.2). At some point, efficiency actually decreases and TPP declines.

The marginal physical product (MPP) is the change in output resulting from a unit change in variable input. The MPP measures the amount that total output changes as inputs are varied and, as such, is the slope of the TPP. The amount of change can be measured by the distance between two points (the way it is usually done in empirical work) or at a point using calculus.

Law of Diminishing Returns

The law of diminishing returns, also called the law of variable proportions, states that if increasing amounts of one input are added to a production process while all other inputs are held constant, the amount of output added per unit of variable input will eventually decrease. The law is ambiguous as to where the point is, but most writers apply the law to the marginal change. For this reason the law is also referred to as the law of diminishing marginal returns.

Three Production Stages

The production function can be divided into three regions or stages (Figure 2.2). TPP in the classical function reflects a situation in which output first increases at an increasing rate, increases then at a decreasing rate, and finally begins to fall. During the first stage (I), in which output is increasing at an increasing rate, APP (the efficiency measure) is continuously increasing. This first stage is also characterized by MPP being greater than APP.

Stage II begins when APP = MPP and ends when MPP = 0, that is, the point at which it becomes negative. The efficiency of using the variable input (as measured by APP) is greatest at the beginning of Stage

II. But the efficiency of the <u>fixed</u> input is greatest at the <u>end</u> of Stage II, the point at which the MPP = 0, because output is maximum for the fixed input. The optimal use of inputs, somewhere in Stage II, depends on input costs and output prices. Stage III is characterized by declining output and a negative MPP. The stages, although delimited by physical factors, provide the transition to determination of optimal output levels from an economic viewpoint.

Optimal Output

The crux of production economics in livestock-systems analysis is that there are three major production decisions that all livestock raisers consciously or subconsciously make, whether they operate under the most primitive nomadic conditions or with extremely sophisticated management. The decisions are (1) how to produce, (2) how much to produce, and (3) what to produce.

The first of the three decisions, "how to produce," provides necessary background information for the determination of what and how much to produce. The decisions are not made just in the formative stages of an operation, but should be periodically reevaluated in light of new information and changing input/output price relationships.

The first step is to set forth a proposed farm plan with alternative enterprises. Then, budgets are developed to provide the economic relations in the "how to produce" decision. After this, an optimal level of production is calculated given economic and management considerations. This is the "how much to produce" decision. Finally, the latter information is compared to determine "what to produce." It would appear that "what to produce" would be the first decision. However, it is last because comparison between alternatives is involved, and that implies first determining the optimal output levels and net income from each one.

The "How" and "How Much" to Produce Decisions

As an example of a "how" to produce decision, first assume that results from the local experiment station show one of the best production methods for steers to be with improved perennial pastures and supplement feed. There are, of course, various supplements, but the research indicates that 16 percent protein concentrate (called input or factor X_1)

has given good results. This factor, along with pasture, which is called X_2, is the factor-factor relationship shown in Figure 2.3. The two factors may be substituted in different proportions and amounts to provide various levels of output, which, represented by the curved lines called isoquants, are denoted as Y_1a, Y_1b, Y_1c in the figure. The lowest output is Y_1a while the highest level of output is Y_1c. The straight lines between the two axes are called isocost lines or constant cost lines as

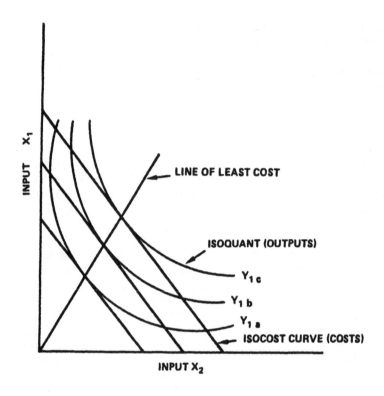

**FIGURE 2.3   EXAMPLE OF THE PRODUCTION ECONOMIC DECISION
ABOUT HOW TO PRODUCE.**

they represent various combinations of inputs that can be purchased by a given outlay.

As greater amounts of the two factors are utilized, higher levels of output, say kilos of beef per steer or hectare, are obtained. There are, of course, an infinite number of isoquants (because there are an infinite number of output levels), just as there are an infinite number of output combinations (along any one isoquant). For simplicity, only two inputs are shown in the figure. The tangency point between an isoquant and isocost line is economically the most efficient (least-cost) combination of inputs to produce that level of output. The tangency points may be connected by a line indicating the least-cost combination of inputs (at the given input prices) to produce any particular level of output. Other combinations of inputs (at the given prices) other than the tangency point are not optimal solutions. The optimal point is where

$$\frac{\Delta Y_2}{\Delta Y_1} = -\frac{P_{y_1}}{P_{y_2}}$$

The points may be connected by a line of least cost that represents the most efficient input combinations for any given set of outputs.

The second economic question, "how much to produce," is graphically shown in Figure 2.4. The vertical line between X_1 and X_2 in the formula $Y_1 = f(X_1 | X_2 ... X_n)$ means that all inputs other than the supplement (X_1) are being held constant. The production function OA shown in the figure bends because output is now represented on the vertical axis while inputs are shown on the horizontal axis. Output increases up to a certain point (C), after which it begins to decline because "too much" of the input is used.

The straight line OB in Figure 2.4 represents the result of dividing the costs associated with the input being varied (X_1) by the price of the output (P_y). In effect, it is $(Cx_1)/(P/y)$. Production textbooks generally show this as $(Px_1)/(P/y)$ under the implicit assumption that there are no other costs. The slope of the line OB (not to be confused with the line OA in Figure 2.2, which is used to define where Stage II begins on TPP) is determined by the relation between this input and output. Optimum output, in this factor-product relationship, is where the MPP is equal to the inverse of the price ratios, that is, the price of supplement divided by price of the output. The equality is determined by the point of tangency. This relationship is also known as producing where marginal revenue just equals marginal cost, where the term marginal means the last additional increment. For example, an "economically rational" producer would continue to increase quantity of supplement until cost of the next input unit just equals the additional income. This assumes that the other input(s), such as pasture (X_2), are held constant at some given

28

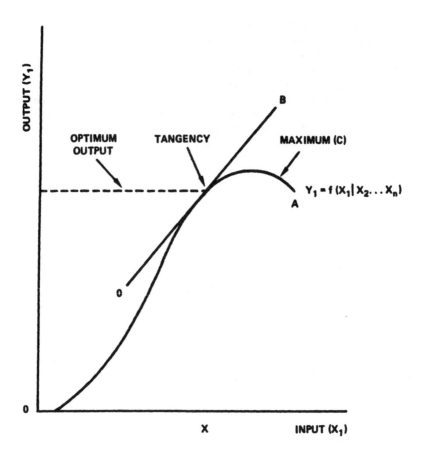

FIGURE 2.4   EXAMPLE OF THE PRODUCTION ECONOMIC DECISION
HOW MUCH TO PRODUCE.

level. The use of these marginal measures is described in the next chapter, at which time an empirical problem is presented.

It is important to recognize that even though the economically optimal production level can be determined, in practice this entrepreneurial management objective is seldom achieved as there are usually some interpersonal, culturally oriented constraints that lead to other input-use levels. For example, although the economically rational operator is assumed to maximize profits, some people feel compelled to maximize production, thus inevitably employing economically excessive input levels. In many cases, quantity of the input used is lower than the economic optimum due to unavailability of supplies or failure to calculate the optimal production level. In some areas, especially Africa, there are social constraints, such as communal grazing systems and status, that lead to greater-than-optimal cattle inventory levels and low offtake (Simpson and McDowell, 1986).

We now digress to an important aspect in the "how to produce" decision (a factor-factor problem). By holding all inputs constant, except the ones being evaluated, it is possible to portray the law of diminishing returns (also called the law of variable proportions) which states that if an input is applied to a fixed factor, eventually, if enough of the variable factor is applied, total output will begin to increase at a decreasing rate (the inflection point) and may approach a maximum, at which time output would begin to decrease (point C in Figure 2.4). In our example, if the supplement is fed at too high levels (relative to an animal's ability to digest it properly), especially in recently weaned calves, many of them probably would become sick and go off feed, and some might even die. The result would be a decrease in output.

The "What to Produce" Decision

The third of the three decisions, "what to produce," is decided upon after sufficient knowledge is obtained about the relevant production alternatives. It is the last of the three decisions because data and analyses about how and how much to produce must first be assembled. Only then can the "what to produce" decision be made by comparing the optimums of the various alternatives.

After the production function information on various alternatives, or at least some estimates, has been obtained, the problem is to relate input costs with output prices. In this relationship, which is graphically depicted in Figure 2.5, the curved lines represent two possible output relationships. For example, a rancher has the option of using his or her land for a dairy operation (Y_1), fattening steers (Y_2), or some combination of both enterprises. There are, of course, many possible output combinations. The amount of potential output depends on the production functions, which, in turn, are related to input use, which, finally, is constrained by resource availability. The curved lines are thus called isoresource or production possibility curves.

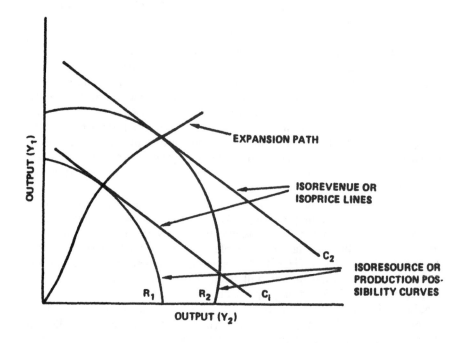

FIGURE 2.5 EXAMPLE OF THE PRODUCTION ECONOMIC DECISION ABOUT WHAT TO PRODUCE.

The optimal amount of each possible output is determined by the tangency of the isorevenue line with the production curves. In other words, the criteria for determining the optimal amount of outputs is based on a combination of the physical, cost and price relationships. If more inputs are used in the production processes, the isocost curve is shifted out, say from C_1 to C_2, so that more of each potential output, such as milk or fed steers, could be produced. The location of C_1 and C_2 is determined by the input constraints, such as the amount of land available.

Only one resource constraint can be shown in Figure 2.5. The tangency of the new isoprice lines with the new isocost curves may or may not lead to the same proportion of outputs. An expansion path can be drawn through the tangency points to show the optimal combinations at different output levels. Given the production possibilities curve associated with the given constraint, the most rational economic decision to maximize net revenue is, as described earlier, to produce at the tangency points where the slope of the isoprice lines (C_1 and C_2) is the inverse of the relative output prices.

The "what to produce" decision in this section is restricted to enterprises Y_1 and Y_2 because that is all that can be shown with the graphs. But many enterprise combinations are possible, and to simplify the very laborious chore of working through each one, a technique called linear programming has been developed. A discussion of it is included in the next chapter. First, a few more concepts must be discussed.

Production Costs

The terms total and average costs are regularly used in the budgeting process. For example, in a cost-and-returns budget for a sheep operation, total annual cost of production can be calculated by summing up each production expense item. This total cost can be divided by the number of lambs sold to calculate an average cost per lamb produced. These total and average costs are for one point on the TPP function. It is not possible to determine from a single-cost budget the stage in which the production is taking place. However, economists do calculate total and average costs of production on existing operations and often develop budgets in a simulation framework to determine the effect of a certain production practice.

Total Costs

It is possible to develop cost estimates for a whole host of production alternatives and to derive cost curves from these estimates. These curves are quite useful both conceptually and empirically for they provide guidelines about the effect from various production and price situations. For simplicity, let us assume once again that all variables except one are held constant. Thus, there is only one variable cost in Figure 2.6. Output is now on the horizontal axis because conventional economic graphical analysis calls for cost to be on the vertical axis.

Total fixed costs (TFC)--that is, those that do not vary during the production period--are a straight horizontal line because they do not change as output changes. Total variable cost (TVC) is computed by multiplying the amount of variable input used by the price per unit of input. Because TVC increases as output increases, the shape of the TVC curve depends on the shape of the production function. For the classical production function, like the one presented in Figure 2.2, TVC will approximate the one in Figure 2.6--TVC will continually increase even though it bends backward to reflect the decline in output after the maximum point on the TPP curve has been passed. Total cost (TC) is a summation of TFC and TVC. This means the distance on the figure between TC and TVC is the amount of TFC.

Average Costs

Average fixed costs (AFC) are determined by dividing total fixed costs by output. Because fixed costs are a constant and production varies, AFC decreases as output levels increase. This tendency is shown in Figure 2.7 where AFC is drawn as a curve that is continuously downward sloping except for a little hook on the right hand corner where it bends back as TPP stops rising and begins to fall.

Average variable cost (AVC) is computed by dividing total variable cost by the corresponding output. The shape of the AVC curve depends on the shape of the production function while the height depends on the unit cost of the variable input. AVC is inversely related to the APP so that when the APP is a maximum, AVC is at a minimum. AVC also begins to bend backward when TPP is at a maximum.

Average total cost (ATC) can be computed by adding AVC and AFC together or by dividing total cost by the corresponding output level. Each ATC computed in the budgets given in the next two chapters represents one point on a curve like the one in Figure 2.7.

Marginal cost (MC) is defined as the change in total cost for a per unit increase in output. In other words, MC is the cost of producing an additional unit of output. Computationally, MC is derived by dividing the change in total cost by the corresponding change in output. As will be shown in the next chapter, MC is the figure, along with marginal revenue, that is used to determine optimal output.

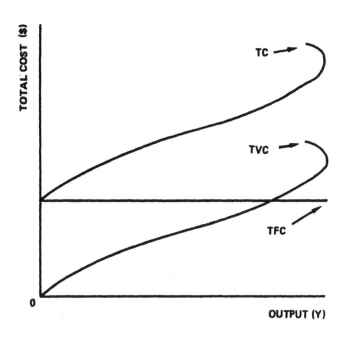

**FIGURE 2.6 COST CURVES FOR A CLASSICAL PRODUCTION FUNC-
TION.**

34

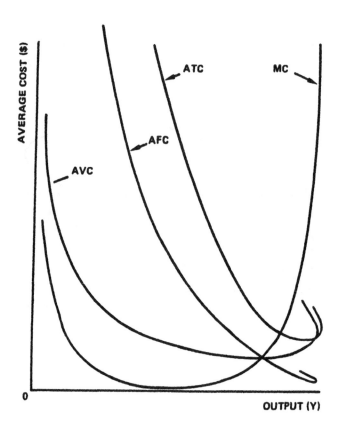

FIGURE 2.7   AVERAGE AND MARGINAL COST CURVES FOR THE CLASSICAL PRODUCTION FUNCTION.

CHAPTER 3

OPTIMIZATION: MARGINAL PRINCIPLES
AND LINEAR PROGRAMMING

The "how much" question in production economics was touched on only briefly in the last chapter to avoid complicating the theoretical principles. Attention now turns to techniques for determining the optimal level of input, output and other resources, that is, the "how much" production decision. The first step is to define terms; then, an example is provided based on two alternative dairy operations, one with purebred cattle (Holstein) and the other with dual purpose (Zebu cross) cattle. The objective is to demonstrate how the optimal level of supplement feed is calculated and also to introduce the budgeting format.

The latter part of this chapter focuses on the method for determining an optimal combination of livestock systems on one farm given constraints on land, labor and capital. The optimal dual-purpose system is then compared with a beef cow/calf enterprise. The objective is to show how, with graphic analysis, linear programming can be used to determine the optimal enterprise combination, that is, the "what to produce" decision.

How Much Input to Use: Marginal Principles

The concept of operating at the margin is related to the "how much to produce" production economics question. It will be recalled a conclusion was reached in the last chapter that profit is maximized by producing in Stage II of the production function. The purpose of the present section is to explain the technique for determining the optimal level of input and output in that stage.

Input Side Decision Rule
The decision rule for determining the optimal input is this: Produce where marginal input cost (MIC) is equal to marginal value product (MVP). MIC is defined as the change in total input cost, or the addition to total input cost caused by using an additional unit of input. MIC is calculated by the equation

$$MIC = \frac{\Delta \text{ Total input cost}}{\Delta \text{ Input level}}$$

MIC is equivalent to the cost of one unit of input, that is, the price of the input. MIC would remain constant provided the input price does not change at higher input-use levels. Care must be taken not to confuse input price changes with volume discounts available to large purchasers of some inputs, which take place regardless of the unit input-use level. If the price does vary, then MIC must use the preceding equation.

MVP is defined as the additional or marginal income received from using an additional input and is calculated using the equation

$$MVP = \frac{\Delta \text{ Total input cost}}{\Delta \text{ Input level}} = MPP \text{ times the output price}$$

Total value product (TVP) is the same as total income or gross income. It is the term used in economics when referring to input levels or use and is simply the total physical product (output) times the selling price.

Output Side Decision Rule

The output side rule is to produce where marginal cost equals marginal revenue (MR), where MC is defined as the change in cost or the additional cost incurred from producing another unit of output. The equation for calculating it is

$$MC = \frac{\Delta \text{ Total input cost}}{\Delta \text{ Total physical product}}$$

or

$$MC = \frac{\text{Price of input}}{MPP}$$

provided price of the input does not change with quantity used.

The second part of the decision rule refers to marginal revenue, which is defined as the change in income, or the additional income received from selling one more unit of output. That equation is

$$MR = \frac{\Delta \text{ Total revenue}}{\Delta \text{ Total physical product}}$$

or

$$MR = \frac{\text{Price of output}}{MPP}$$

Total revenue is the term used in economics to mean total income, when operating on the output side. In effect, total revenue is the term used in place of total value product, which is an input side term. MR, which is the same as the output price, will virtually always be constant at different production levels because farmers very seldom influence market price by changing their production levels. If the selling price does vary, then MR must be calculated using the first equation.

Both decision rules provide the same results. The major differences are that in MIC = MVP the input price is held constant, while in MC = MR output price is held constant.

Decision Rule Application

Information on cattle inventory, number of cattle marketed, various production measures and prices for an example problem is provided in Table 3.1. With that basic information having been assembled, the next step is to calculate optimal supplemental feed use to demonstrate the use of marginal analysis.

The daily per head milk production (or TPP) for the Holstein cattle at various supplement levels is provided in Table 3.2. In addition, calculations of MPP, MVP and MIC are also provided. The computations indicate that when 0.90 kg of supplement is fed daily, a milk production level of 5.82 kg per cow can be expected. The additional output, or MPP with respect to the previous supplement level, is 0.90 kg. Price of milk is $0.37 per kg so MVP, the result of multiplying MPP times that price, is $0.33 when moving from 0.45 to 0.90 kg daily of supplement. Marginal input cost, or MIC, is equivalent to the price of supplement ($0.26) and is less than the MVP ($0.33), thereby indicating that a higher supplement level is optimal. The optimal input-use level is 2.73 kg because moving to the next higher level (3.20 kg) would result in the additional return (MVP = $0.22) being lower than the input cost ($0.26).

The last two columns of Table 3.2 contain the results of calculating MC and MR. Utilization of the alternative decision rule to produce when MC = MR yields the same result--that is, 2.73 kg is the optimal feed level. An input level beyond that would result in the additional (marginal) cost being greater than the additional revenue. For example, going from 2.73 kg to 3.20 kg would result in MC increasing from $0.35 to $0.43 per kg, which exceeds the MR of $0.37.

Table 3.1. Basic data for example dairy and cow/calf operations

Item	Units	Purebred dairy	Dual purpose	Beef cow/calf
Inventory				
Total animal units	AU	100	100	100
Pasture land	ha	100	100	100
Cropland	ha	-	-	-
Livestock				
Mature cows	no	72	83	85
Heifers	no	25	15	11
Yearling bulls	no	1	1	1
Bulls	no	3	3	3
Horses	no	-	-	1
Total animals	no	101	102	101
No. of animals marketed annually				
Cull cows	no	24	12	10
Cull bulls	no	1	1	1
Heifer calves	no	8	20	22
Steer or bull calves	no	32	32	33
Total animals	no	65	65	66
Production measures				
Calf crop (weaned)	pct	90	90	80
Death loss (weaned calves and older)	pct	1	1	1
Replacement rate, females	pct	33	17	12
Milk production per cow (annual)	kg	2,681	1,150	-
Total milk production	kg	193,032	95,438	-
Sale weights (shrunk)				
Cows	kg	455	425	400
Bulls	kg	600	575	550
Heifer calves	kg	35	40	180
Steer calves	kg	35	40	190
Total calf sale weight	kg	1,400	2,080	10,230
Total cattle sale weight	kg	12,920	7,755	14,780
Beef production per ha	kg	129	78	148
Milk production per ha	kg	1,930	954	-
Prices				
Calves	U.S.$/kg	0.80	0.80	0.95
Cull cows	U.S.$/kg	0.75	0.75	0.75
Cull bulls	U.S.$/kg	0.85	0.85	0.85
Milk	U.S.$/kg	0.37	0.37	-
Supplement feed (16% protein)	U.S.$/kg	0.26	0.26	-

39

Table 3.2. Determination of optimal supplement feed level for purebred Holstein cows, example problem

Daily per head ration of supplement	Daily per head milk production TPP	Marginal physical product MPP	Marginal value product MVP	Marginal input cost MIC	Marginal revenue MR	Marginal cost MC
	Kilos			U.S.$		
0.00	3.87	1.05	0.39	0.26	0.37	0.25
0.45	4.92	0.90	0.33	0.26	0.37	0.29
0.90	5.82	0.74	0.27	0.26	0.37	0.35
1.35	6.56	0.74	0.27	0.26	0.37	0.35
1.80	7.30	0.75	0.28	0.26	0.37	0.36
2.25	8.05	0.74	0.27	0.26	0.37	0.35
2.73	8.79	0.60	0.22	0.26	0.37	0.43
3.20	9.39	0.59	0.22	0.26	0.37	0.44
3.65	9.98	0.22	0.08	0.26	0.37	1.18
4.10	10.20	0.30	0.11	0.26	0.37	0.87
4.55	10.55					

MVP = MPP times output price.
Price of supplement (input) = $0.26.

MC = Price of input divided by MPP.
Price of output (milk) = $0.37.

Purebred Dairy Breed Costs and Returns

The data on inventory and production measures for a purebred (Holstein) alternative are given in Table 3.1. The major constraining factor, as in the dual-purpose cattle example just given, and in a beef breed cow/calf alternative to be presented, is that only 100 hectares of land are available; however, a small part of it is seeded and fertilized to provide more and better quality forage for milking cows. The carrying capacity is limited to 100-102 AUs. Thus, fewer purebred cows (72) can be carried than in the case of the dual-purpose breed (83) because more replacements are raised (cows are culled every three years for the Holstein operation versus every six years for the dual-purpose breed). A 90 percent calf crop and 1 percent death loss are assumed in each case.

Introduction to Budgeting

A complete cost-and-returns budget for the purebred operation as well as dual-purpose and beef cattle is provided in Table 3.3. Detailed explanation of the budgeting procedure is deferred until the next chapter to avoid complicating the discussion. However, an explanation of the categories themselves is now provided as an introduction to the specialized budgets for dairy and beef cow operations.

The first category in Table 3.3 is investment, which is composed of land, buildings and other capital investments such as fences, equipment and livestock. The total for the purebred operation is $128,400, mostly in land and livestock.

Annual production costs are divided into two parts, basic production costs and other costs. The basic production costs are ones that will nearly always involve a cash outlay, while items included in "other" may or may not involve cash. For example, in the category "capital costs," a charge might be incurred for interest on a loan or an opportunity cost for the money invested. Opportunity cost is defined as the value of the product not produced because an input was used for another purpose. Opportunity cost is an economic concept that recognizes that once an input is committed to a particular use, the input is no longer available for any other alternative, and, as such, the income from the alternative must be foregone.

There is no charge in Table 3.3 for operating capital in the dairy operations because milk is delivered daily and payment is assumed to be received on a regular, relatively short basis. Thus, the producer need not tie up a lot of operating capital as happens in cow/calf operations where cattle are typically only sold once a year. If payment for milk were regularly delayed, then a charge on operating capital would be included.

Depreciation is a legitimate charge, even though it is not a cash cost, because at some point buildings and equipment will have to be replaced. The charge is thus for something wearing out. Care must be taken to not confuse this term with amortization, which is a periodic payment such as the gradual extinguishment of a mortgage. Taxes and any insurance are cash payments but are classed as ownership costs because they are not directly related to the production process per se.

Personal labor is classed in other costs because producers will seldom pay themselves a regular salary or hourly wage. Rather, they are residual claimants to profits even though living expenses may be withdrawn from the firm during the production process. The same holds true for management.

Costs and Returns

Total basic production costs amount to $38,804 for the purebred operation. Part of that total is supplemental feed for basic maintenance of cattle such as replacement heifers, bulls, dry cows, and cows on pasture during periods of the year. Added to that total cost is the production enhancing supplement feed cost of $15,594, which is the product of multiplying the optimal daily supplement level of 2.73 kg (determined in Table 3.2) times 305 days of lactation to arrive at total annual use per head (833 kg). That product is multiplied by 72 cows to calculate total additional supplement use (59,976 kg) and then times $0.26 per kg.

The next step in Table 3.3 is to subtract salvage income from sale of cull cows and bulls to determine net basic production cost. This procedure is necessary to determine the breakeven cost, that is, the cost of production.

Other costs, estimated at $32,794, include capital costs, ownership costs, personal labor and a management charge. When other costs are added to net basic production costs the total is $78,492.

Income from calves is estimated at $1,120 annually while milk revenue is $71,422. Net income above production costs is $26,848 while it is a negative $5,950 above all costs.

There is no cost of production estimate for calves because the main objective is to calculate milk production costs. That latter estimate entails subtraction of calf income ($1,120) from net basic production costs and total costs and then division of the remainder ($53,278 and $77,372) by annual milk production (193,032 kg). Milk production is the product of multiplying the daily per head milk production of 8.79 kg (Table 3.2) times 305 days of lactation (2,681 kg per cow) times 72 cows. The net result is a milk production cost of $0.28 per kg if only net basic production costs are considered and $0.40 per kg when all costs are taken into account.

Dual-Purpose Breed Costs and Returns

Costs and returns for the dual-purpose system are calculated in the same manner as those just presented.[1] Thus, the first step is to use the marginal principles to determine optimal supplement feed level for milk production enhancement. Because dual-purpose cattle such as a Shorthorn-Zebu or Jersey-Zebu cross do not respond as well to supplement

Table 3.3. Annual costs and returns for 100 animal unit
purebred dairy, dual purpose and beef cow/calf
operations example problem

Item	Purebred dairy	Dual purpose	Beef cow/calf
	- - - - - U.S.$ - - - - -		
Investment			
Land	50,000	50,000	50,000
Buildings and capital investment	15,000	12,000	2,000
Equipment	7,000	5,000	3,000
Livestock	56,400	45,100	35,500
Total	128,400	112,100	90,500
Basic production costs			
Hired labor	12,440	10,080	700
Supplemental feed maintenance	12,220	6,610	5,790
Salt and minerals	475	475	475
Repairs and maintenance			
Buildings and improvements			
	1,575	100	50
Machinery and equipment	1,894	50	50
Veterinary supplies and services	700	400	400
Seed and fertilizer	4,000	3,000	3,000
Bulk and/or artificial			
insemination	4,000	1,500	1,000
Machinery			
Operating expenses	400	50	350
Hired	300	250	-
Transportation	100	100	75
Utilities	400	50	-
Other and administrative	300	200	100
Subtotal	38,804	22,865	11,990
Supplemental feed, increased			
production	15,594	2,962	-
Total	54,398	25,827	11,990
Less salvage income			
Cull cows	8,190	3,825	3,000
Cull bulls	510	489	468
Total	8,700	4,314	3,468
Net basic production cost	45,698	21,513	8,522

continued

Table 3.3. (continued)

Item	Purebred dairy	Dual purpose	Beef cow/calf
	- - - - - U.S.$ - - - - -		
Other costs			
Capital costs			
Land	8,000	8,000	8,000
Buildings and capital investment	2,400	1,920	320
Equipment	1,120	800	480
Livestock	9,024	7,216	5,680
Operating capital	-	-	539
Subtotal	20,544	17,936	15,019
Ownership costs			
Depreciation	1,450	1,100	400
Taxes	400	375	300
Insurance	400	50	-
Subtotal	2,250	1,525	700
Labor, own	-	-	-
Management	10,000	8,000	4,000
Total, other costs	32,794	27,461	19,719
Total, net basic production and other costs	78,492	48,974	28,241
Income			
Steer calves	896	1,024	5,957
Heifer calves	224	640	3,762
Subtotal	1,120	1,664	9,719
Milk	71,422	35,312	-
Total	72,542	36,976	9,719
Net income			
Above net basic production costs	26,848	15,463	1,197
Above all costs	-5,950	-11,998	-18,522
Cost of production calves			
Net basic production costs	-	-	0.83
All costs	-	-	2.76
Cost of production, milk[a]			
Net basic production costs	0.28	0.23	-
All costs	0.40	0.51	-

[a] Income from calves is subtracted from costs.

feed as purebreds do, only the lowest level of supplement feed, 0.45 kg, is found to be optimal (Table 3.4). A move to the next higher level of 0.90 kg means that MVP would fall to $0.24 or $0.02 below MIC.

Total cost of the additional supplemental feed is only $2,962. Consequently, net basic production costs at $25,827 are less than one-half those for the purebred operator. Other costs are calculated to be $27,461; thus, total annual costs are $48,974.

On the income side, milk output of 3.77 kg per day per cow in production (Table 3.4) results in a total annual milk production of 95,438 kg, which at $0.37 per kg is $35,312 (Table 3.3). That, added to $1,644 from calves, is a total income of $36,976. Net income above basic production costs is $15,463, but there is a loss of $11,998 when all costs are considered. Milk production cost is $0.23 per kg when only net basic production costs are considered and $0.51 per kg when all costs are taken into account.

Discussion

The losses incurred when all costs are considered are typical of cattle operations throughout the world. If the major source of that loss--opportunity cost on capital invested--were deleted from the analysis, the result quite often would be reversed. But budgeting will reveal that in very competitive situations many producers will not receive any return to management and in some extreme cases may even subsidize their cattle operations from other enterprises. Analysts should realize that their budgeting is correct and instead of changing assumptions as might be the first reaction, should look for reasons behind the results.

Setting up the categories in the manner shown in Table 3.3 is very useful for it provides a means whereby the economic relationships can be disaggregated and analyzed. The large differences between basic production costs and total cost are also clear indicators of the need for extreme caution in carefully specifying the items included in the estimates.

A major objective of this section has been to present the dairy and beef cow budgeting format, to show calculation of production cost, and to indicate how to apply marginal principles in the "how much" production decision question. In addition, the budgets can also serve as a framework to identify optimal breeds (Harris, Steward and Arboleda, 1984). Furthermore, even though the economic relationships are only for an example problem, they are representative of differences between dual-purpose and Holstein cattle in many countries (Salmon and Warnken, 1982). More total profit and profit per cow per hectare can be obtained from the purebreds provided management, capital and adequate output price are available. That, of course, is a big if, and lack of one or more of these factors is the reason why so-called dual-purpose cattle have received so much attention in tropical and subtropical areas.

Risk reduction is another very good reason for extensive use of dual-purpose-type cattle in developing countries. As an example consider a

Table 3.4. Determination of optimal supplement feed level for dual purpose cows, example problem

Daily per head ration of supplement	Daily per head milk production TPP	Marginal physical product MPP	Marginal value product MVP	Marginal input cost MIC	Marginal revenue MR	Marginal cost MC
	- - Kilos - -		- - - - - - - -	- - U.S.$	- - - - - - - - -	- - - -
0.00	2.95					
0.45	3.77	0.82	0.30	0.26	0.37	0.32
0.90	4.43	0.66	0.24	0.26	0.37	0.39
1.35	5.03	0.60	0.22	0.26	0.37	0.43
1.80	5.66	0.63	0.23	0.26	0.37	0.41
2.25	6.23	0.57	0.21	0.26	0.37	0.45
2.73	6.73	0.49	0.18	0.26	0.37	0.53
3.20	6.97	0.25	0.09	0.26	0.37	1.04
3.65	7.18	0.21	0.08	0.26	0.37	1.23
4.10	7.34	0.16	0.06	0.26	0.37	1.63
4.55	7.38	0.04	0.01	0.26	0.37	6.50

MVP = MPP times output price.
MC = Price of input divided by MPP.

Price of supplement (input) = $0.26.
Price of output (milk) = $0.37.

producer in a developing country with 10-30 head of cows. If children leave home or a reliable milker cannot be found, the owner can simply stop selling milk and just let calves suckle. In fact, calves are usually allowed to suckle on most dual-purpose operations in developing countries as a complement to milking. Producers with 10-30 head can be considered commercial operators even though they produce on a relatively small scale. Milk marketed through calves can be valued either on an opportunity cost basis by estimating the quantity fed or by including calves as an income item from the operation as was done in Table 3.3.

The data in Table 3.1 show there is only slightly more calf sale weight in the dual-purpose breed and about one-third less total animal weight when all cattle sales are accounted for. It is true that having dual-purpose cattle based on a beef breed may provide more beef per hectare from the growing and fattening process in raising steers, but that is a completely different enterprise and, consequently, must be budgeted separately.

Dual-purpose cattle form an integral part of livestock systems in some regions because producer returns can be maximized from dual purpose cattle given market and policy situations at a particular point in time. Zebu-based dual-purpose cattle are found in much of the developing world because they serve as a "low management survival breed." Another reason why purebred milk cow systems have not developed to an even greater extent than they have in less-developed countries (LDCs) is that the 1970s and early 1980s constituted a period of attention on small rather than large operators and on propagation of a hardy breed that would survive with low management and little use of inputs. Even more important is the extremely large volume of heavily subsidized powdered milk and milk products that has become available on the world market. Governments in developing countries, ever mindful of consumer pressure, have found imported products to be cheaper, especially in the shorter run, than developing national dairy industries. Overall, the examples not only demonstrate the usefulness of a systems approach, but also the need to carefully consider macroeconomic as well as microeconomic factors in analyses.

A Beef Cow/Calf Operation as a Last Example

The alternative of using the 100 ha for a beef breed cow/calf operation rather than dairy cattle is now budgeted. Stocking rate is still 1 animal unit per hectare, which means that the land has a capacity for 85 mature cows. There is an 80 percent calf crop and 12 percent replacement rate. Thus, 22 heifer calves and 33 male calves are sold annually (Table 3.1). With steer calves marketed at 190 kg and heifer calves at 180 kg, there is a production of 148 kg per hectare (including cull animals). In contrast, the Holstein operation yields 129 kg of live animal sales per hectare, while the dual-purpose cattle yield 78 kg per hectare. The owner could elect to fatten the calves on another piece of land, in which case the sale price is an accounting transfer charge from one enterprise to another rather than cash actually changing hands.

Net basic production expenses amount to $8,522 while other costs add another $19,719, for a total of $28,241 (Table 3.3). Gross income is $9,719. Net income above net basic production costs is $1,197, while a loss of $18,522 is incurred when all costs are accounted for.

Cost of production is $0.83 per kg when only net basic production costs are considered and $2.76 when all costs are accounted for. As with the previous budgets, the analyst can delete items in the other costs category and recalculate production cost according to the problem at hand. In effect, the interpretation is that cost of production lies between $0.83 and $2.76 per kg depending on the items included. This example highlights the need to carefully specify items to be included in cost estimates.

The input/output relationships in the examples have been derived from actual data, but no attempt is made to recommend any one alternative. Rather, the intention is to show the method for determining the optimal system under given conditions. Simple budgeting is adequate here as an analytical tool because there are no constraints, such as labor or capital, placed on the operation, which would make a combination of enterprises more profitable than a single enterprise. The problem of multiple constraints is dealt with in the next section.

Linear Programming in Livestock Production Economics

The previous examples contained an implicit assumption that the optimal combination of supplemental feed (called input X_1) and pasture or roughage (X_2) had been determined and that the resulting production function only had one variable input, X_1. But there are many other combinations of inputs that could have been used. As might well be imagined, the arithmetic can quickly become tedious when an attempt is made to find an optimal solution to a problem involving two or more inputs in conjunction with two or more products. As a consequence, a quantitative method called linear programming (LP) was designed to handle this task. The method was greatly popularized by Earl O. Heady and Wilfred Candler in their 1958 book entitled Linear Programming Methods and has become one of the most widely utilized tools in agricultural economics as well as in many other disciplines. Linear programming can be used with budgets like those shown in Table 3.3, but it is not a substitute for budgeting. Space is devoted to explaining the basics of linear programming because of the close tie to budgeting, LP's usefulness in investment analysis, and the method's theoretical value in conceptualizing livestock investment and development problems. For these reasons, the discussion of LP in this book is restricted to conceptual aspects of LP in livestock analysis; virtually no attention is given to the mechanics of application.

An important aspect of linear programming, and the reason for its name, is that the production relationships are assumed to be linear. In other words, in linear programming each of the inputs is utilized in fixed proportions, which means that output is determined by the limiting input. This is because in LP one input cannot substitute for another one. For example, in an LP problem, within one activity, machinery cannot be used in place of labor. Rather, the two are used in a fixed ratio to each other. However, the two could be interchanged between activities, such as hay or pasture.

Care must be taken to avoid confusing input use as defined in linear programming with the production functions described in Chapter 2. In the production function all inputs are held constant except one, which, if increased enough, would lead output to diminish. In contrast, input combinations in LP constitute the one input, for in linear programming all inputs increase at a fixed rate. Output would, of course, continue to expand to unreasonable extremes in linear programming problems as a result of fixed input-output relationships if it were not for the restrictions on the amount of inputs available. These limitations, called constraints, constitute the first of three steps in setting up a linear programming problem. These steps are

1. Define the constraints;
2. Develop the objective function, which is either maximization or minimization of something;
3. Set forth the alternative ways to achieve the objectives.

Let us consider LP concepts by working through a simple maximization problem that is an expansion of the example presented earlier. Suppose that the producer has the 100 hectares previously described, but only has access to $10,000 in operating capital for production expenses and just 200 hours of labor (Table 3.5). These are the constraints. Also suppose this person only wants to use the land for cattle (no crops) and that the options being considered are a cow/calf breeding herd, a dual-purpose dairy operation, or a combination of the two enterprises.

Our objective is to determine the optimal amount of land for each operation. In other words, we are once again dealing with the "what to produce" part of production economics. The input costs, labor, and operating capital requirements for the enterprises are given in Table 3.5 along with the outputs and expected prices from each type of operation. Note that all of the specifications are the same as those assumed in the earlier examples.

This LP problem can be solved geometrically by labeling hectares devoted to the dual-purpose operation on the horizontal axis and hectares in the cow/calf enterprise on the vertical axis (Panel A, Figure 3.1). The second step is to plot the three constraints. In Panel A, a point is marked at 100 hectares of land on both the vertical and horizontal axis, and a straight line is drawn between them. This line reveals all possible

Table 3.5. Input and output specifications for linear
            programming example

Input or output	Input availability	Dairy, dual purpose	Cow/calf
Beef production per hectare calves (kg)[a]	-	20.8	102.3
Production costs per hectare ($)[b]	-	215	85.22
Milk production per hectare (kg)[a]	-	954	-
Price per kilo			
Beef ($)[c]	-	0.80	0.95
Milk ($)[c]	-	0.37	-
Labor			
Available (hr)	200	-	-
Required (hr/ha)	-	10.0	1.4
Hectare constraint (ha)[d]	-	20	142
Operating capital			
Available	10,000	-	-
Required ($/ha)	-	70	80
Hectare constraint (ha)[d]	-	143	125

[a]See Table 3.1.
[b]See Table 3.1.
[c]See Table 3.3. Net basic production costs divided by 100.
[d]Product of dividing amount available by amount required.

combinations in land use between each enterprise. The cattleman could
use less than 100 hectares because land is not the only restriction, but
the maximum is 100 hectares.

The next step is to add the labor constraint (Panel B), which already
has the land constraint drawn in. A maximum of 200 hours of labor is
available, which means only 20 hectares could be devoted to the dual-
purpose operation. Because only 1.4 hours per hectare are needed for the
cow/calf enterprise, 142 hectares could potentially be handled. As with
the land constraint the two points are plotted on the vertical and
horizontal axis, and a straight line is drawn between them. This line
shows all possible combinations of dual-purpose and cow/calf operations
that can be produced with only 200 hours of labor.

The capital constraint is added in after labor. Calculations indicate
that the $10,000 of available operating capital could permit use of 125
hectares ($10,000 ÷ $80/ha) in a cow/calf operation. Application of all
the capital for a dual-purpose operation would permit 143 hectares
($10,000 ÷ $70/ha) to be utilized if that much land were available. Each
of these points is plotted on the appropriate axes in Panel C, to which
the land and labor constraints have previously been drawn in. The

capital constraint line is outside the land and labor constraint lines, which means that capital is not a limiting factor in either one of the operations.

Now that all of the constraints have been properly plotted, the profit maximization combination can be identified. The optimal level will always be located at a "corner" where the inside constraint lines intersect or where the inside constraint lines cross the axis. These points are given the labels O, A, B and C in Panel D, which has been redrawn from Panel C. Land and labor are the two effective constraints. The optimal production combination will never fall on the straight line segment of the restriction lines.[2]

One way to determine the most profitable "corner" is to calculate income and cost at each corner, as in Table 3.6. In corner O, which is the intersection of the cow/calf and dual purpose axes, there is no production and consequently no net income. Corner A, which is constrained by land at 100 ha, is budgeted as if all resources were used in the cow/calf operation. The 100 hectares given in this activity solution are multiplied by the originally specified 102.3 kilos of live beef production per ha, which at $.95 per kilo provides a gross income of $9,719. The operating cost of $85.22 per ha results in $8,522 for the 100 hectares or a total net income of $1,197 for the entire operation. This is the same net income as shown in the enterprise budget developed in Table 3.3. The consistency between budgeting and LP has also been demonstrated by Kottke (1961).

Corner B is the intersection of the land and labor constraints. The optimum for the cow/calf operation is 92.5 hectares with the other 7.5 hectares devoted to the dual purpose enterprise. Net incomes of $1,107 and $1,160 are derived from the two operations respectively, for a total net income of $2,267. Corner C, in which the only enterprise is dual purpose cattle, provides a net income of $3,092, even though only 20 hectares can be utilized given the constraints. This same solution can be obtained by multiplying the net return per hectare ($15,463 ÷ 100 = 154.63 in Table 3.3). The conclusion is that with the given constraints, the landowner would be best off with only 20 hectares of dual-purpose cattle, leaving the rest of the land idle.

It can be argued that in most developing countries labor availability is not a problem. But that constraint can be quite realistic if one considers management as part of labor. For example, it may be that an owner does not have time to manage, at the level specified, more than 200 hours of labor.

The optimal solution incorporates the answers to the three production questions "how," "how much," and "what" to produce. The "what" analysis indicates that a strictly dual-purpose operation, even though all land is not used, would maximize income. The "how" and "how much" questions are included in the production cost and output specifications. The analysis shows that some land and capital would be left over so that if the operator had correctly specified the production function relationships,

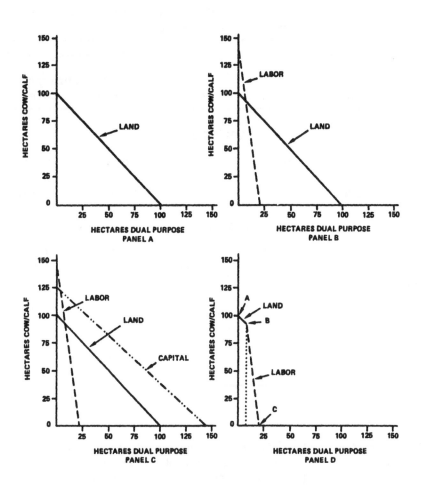

FIGURE 3.1 DETERMINIATION OF CORNER SOLUTIONS, LINEAR
PROGRAMMING EXAMPLE.

Table 3.6. Calculations of net returns from various corner solutions, linear programming example

		Cow/calf			
Item	Land	Calf production per ha	Operating cost per ha	Price per kilo	Enterprise total
	-Ha-	- Kg -	- - - - - - - U.S.$ - - - - - -		

<u>Corner A</u>

Item	Land	Calf production per ha	Operating cost per ha	Price per kilo	Enterprise total
Income	100.0	102.3	-	0.95	9,719
Cost	100.0	-	85.22	-	8,522
Net	100.0	-	-	-	1,197

<u>Corner B</u>

Item	Land	Calf production per ha	Operating cost per ha	Price per kilo	Enterprise total
Income	92.5	100.3	-	95	8,990
Cost	92.5	-	85.22	-	7,883
Net	92.5	-	-	-	1,107

<u>Corner C</u>

Item	Land	Calf production per ha	Operating cost per ha	Price per kilo	Enterprise total
Income	0	0	0	0	0
Cost	0	0	0	0	0
Net	0	0	0	0	0

continued

then 80 hectares and $8,600 in operating capital could be invested in other activities. In other words, the analysis not only provides an answer to the three questions, but also provides information that can be used for planning-related operations.

The linear programming example provides a framework to understand the technique and how it is used. Naturally, most LP problems have many more activities and thus are solved with computers rather than by hand. In addition to the maximization type problem, a common use of LP is in least-cost cattle-ration formulation and cost minimization in transportation problems. Also, LP has been extended to regional and national problems for whole sectors (Simpson and Farris, 1982). Other topics of special interest to researchers on livestock systems have been applications of LP to dairy cow replacement (Smith, 1971), effects of forage quality restrictions on optimal production systems (Whitson, Parks and Herd, 1976), and generalized use for beef cattle production problems (Wilton et al., 1974). As shown by Miller, Brinks and Sutherland (1978) the computer will be a major, if not essential, tool in management decision making for beef production systems in the future.

Table 3.6. Calculations of net returns from various corner
solutions, linear programming example (continued)

		Dual purpose				
				Operating		
		Production per ha		cost	Price per kilo	
Item	Land	Beef	Milk	per ha	Calves	Milk
	-Ha-	- - - Kg - - -		- - - - - U.S.$ - - - - -		

Corner A

Item	Land	Beef	Milk	Operating cost per ha	Calves	Milk
Income	0	0	0	0	0	0
Cost	0	0	0	0	0	0
Net	0	0	0	0	0	0

Corner B

Item	Land	Beef	Milk	Operating cost per ha	Calves	Milk
Income	7.5	20.8	954	-	0.80	0.37
Cost	7.5	-	-	215	-	-
Net	7.5	-	-	-	-	-

Corner C

Item	Land	Beef	Milk	Operating cost per ha	Calves	Milk
Income	20.0	20.8	954	-	0.80	0.37
Cost	20.0	-	-	215	-	-
Net	20.0	-	-	-	-	-

Item	Enterprise total	Total both enterprise
	- - - - - - U.S.$ - - - - - -	

Corner A

Item	Enterprise total	Total both enterprise
Income	0	-
Cost	0	-
Net	0	1,197

Corner B

Item	Enterprise total	Total both enterprise
Income	2,773	11,763
Cost	1,613	9,496
Net	1,160	2,267

Corner C

Item	Enterprise total	Total both enterprise
Income	7,392	-
Cost	4,300	-
Net	3,092	3.092

NOTES

1. This section draws heavily on information contained in Calo et al. (1973); de Alba (1978); Frisch and Vercoe (1978); and Simpson (1982).

2. This is not exactly correct because in this case where the optimal solution can be either B or C (indeterminate solution) when any points on the line between B and C would be optimal. However, LP algorithms will always find the corner solutions.

CHAPTER 4

CATTLE SYSTEMS IN WHOLE-FARM ANALYSIS

The basic outline for livestock budgets and the method to determine optimal input levels were provided in the last chapter. The purposes of the present chapter are (1) to show how the basic budget format can be modified to fit a variety of different systems on one farm and (2) to explain how livestock enterprises can be evaluated within a whole farm planning context. Another interpretation of the examples is the way to describe cost-and-returns analyses on livestock systems in a certain geographic area.

The budget formats provided in this chapter are especially relevant to the large commercial operations found in much of Latin America. But the format also applies to the smallest subsistence-level operation, such as an operation with just one cow or buffalo for draft, even though few of the cells will have entries. In large part, this chapter's value for subsistence-type operations with virtually no purchased inputs may be as a tool for concept development rather than for actual quantitative analysis. The examples have been chosen because the operations extensively use purchased inputs, which permits exposition of a budget with a full complement of components. Budgets of this type are widely applicable to the voluminous number of commercial (small as well as large) operations found around the world. Budgeting on very small and subsistence-level operations is discussed in Chapter 9.

Whole-Farm Planning

Several livestock enterprises may be found on one farm or ranch, while in other cases, such as a cow/calf ranch in Paraguay or a nomadic herder in the Sahael of Africa, the entire operation may be composed of just one enterprise. Regardless of the operation's sophistication or simplicity, the first step in producer-level analysis is to describe the resources available and to develop one or several whole-farm plans.

If the operation is relatively sophisticated, various records about operational details may be required. But if the operation is at the subsistence level, detailed records will not be required, although the analyst must be equally cautious to account for all activities and resources. An apparently innocuous piece of wood may be just as important to a desert herder as sophisticated purchased machinery is to a European farmer.

Once the physical resources are categorized and the organizational scheme or systems are identified, the expected cost and returns for the plan can be organized into a whole farm budget. The combined result is a detailed physical and economic plan for organization and operation.

A major objective of this book is to show how livestock investments can be analyzed rather than to simply present farm management principles. Consequently, the important steps of resource identification and classification and preparation of whole farm plans that are well documented in standard farm management textbooks are bypassed. Discussion thus proceeds directly to budgeting alternatives for various types of beef cattle systems. The numbers are realistic for many farm situations in the late 1980s. But relatively little attention should be given to the numbers because the focus of the chapter is on concepts. In effect, the data are glue rather than primary material.

Description of the Operation

The example farm used throughout this chapter has a portion in crops and 100 hectares in pasture with a carrying capacity of 1 ha per mature cow unit. It is further assumed that part of the year cows can be grazed on corn stubble, in timber or on other parts of the farm. Hay is harvested each summer from 35 ha, and a cool season annual grass (rye) is oversown on 9.7 ha each winter (cold period).

The alternatives evaluated are cow/calf, weaning at 205 kg.; growing from 205 to 320 kg.; grain finishing from 320 to 461 kg.; forage finishing from 320 to 465 kg.; combination forage/grain finishing from 320 to 470 kg. A schedule of activities, in terms of time requirements, is given in Figure 4.1.

Weaning at 205 Kilos

Cattle

The budget for weaning steers at 205 kg assumes the following inventory:

Item	Number	Animal unit equivalents	Animal units
Mature cows	100	1.0	100
Replacement heifers	25	0.7	18
Bulls	5	1.2	6
Total	130		124

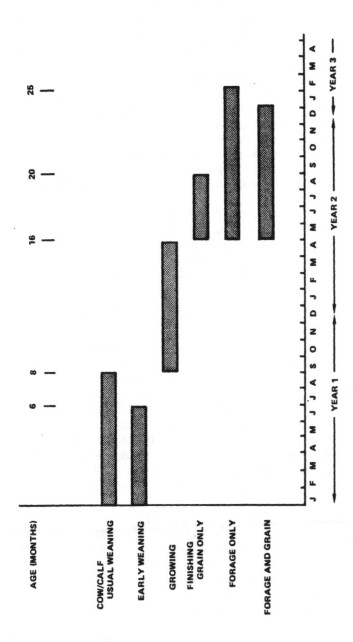

FIGURE 4.1   SCHEDULE OF ACTIVITIES, EXAMPLE CATTLE OPERATION.

Further assumptions are that an 85 percent calf crop is achieved and that one of the replacement heifers will die or will be culled, so there are 24 first-calf heifers. The heifers are bred at 15 months of age and thus begin calving at 24 months. Also, there are annual sales of 23 cull cows (allowing for a death loss of 1 cow) weighing 405 kg each. One 675 kg bull is replaced each year. Cattle are commercial grade crossbreds. The herd consists of a total of 130 head, which represents 124 animal units, where an animal unit is the equivalent of one mature cow.

A budget for a cow/calf operation begun in the early 1980s is given in Table 4.1. The footnotes should be referred to because the objective in this chapter is to describe how numbers are developed rather than to concentrate on the numbers themselves. It is very important that the person developing budgets keep detailed notes on calculations, whether they be done on a computer or by hand, so that changes can easily be made in response to price fluctuations or management plan adjustments.

Investment, Feeding Program, and Production Costs

It is determined that breeding animals must be fed a supplement from November until April to meet forage deficiencies due to cold weather. A typical supplemental feeding program by animal type is graphically portrayed in Figure 4.2; details on quantities are provided in Table 4.2. The basic program consists of hay produced as part of the cattle operation. A total of 141 tons of hay is fed along with 11.2 tons of range cubes (a pressed protein-enriched supplement feed) and 9.1 tons of grain. In addition, there are 9.7 ha planted annually with a rye and ryegrass mixture for the replacement heifers (December 1 to February 15), and for the first-calf cows (February 15 to May).

Total investment for the operation is $265,879. The term basic production costs, as explained in the last chapter, is used instead of the term cash costs because several items such as labor, owner's management charge and taxes are shown separately under "other costs." Production costs for the hay are included in Table 4.1 as are fertilizer, nitrogen, equipment maintenance, fuel and repairs. There is a separate category for hay harvesting costs under the assumption that a charge is made to the cattle operation by the farm owner or by an outside operator.

Other Costs, Net Income and Breakeven Analysis

In addition to the basic production costs, there are a number of other expenses that must also be considered, such as capital costs, (those in which there may be a direct cash outlay if the money is borrowed) or an opportunity cost if personal capital is tied up in this enterprise rather than being invested in an alternative operation. The capital costs, amounting to $15,533, are calculated using 12 percent as the cost factor. Ownership costs, which include depreciation, taxes and insurance, add another $8,858. The taxes and insurance are direct cash outlays while cash may or may not be retained for depreciation.

59

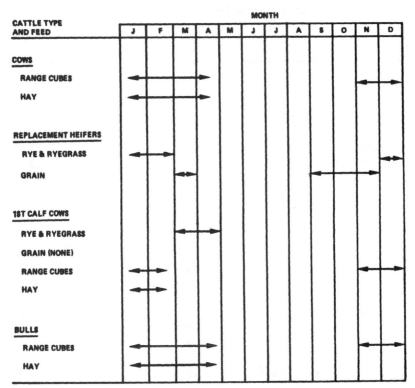

ᵃ SEE ALSO TABLE 4.2 FOR QUANTITIES OF SUPPLEMENTAL FEED.

FIGURE 4.2   SUPPLEMENTAL FEEDING PROGRAM, EXAMPLE COW/-
CALF OPERATION.

Table 4.1. Costs and returns for example 100-brood-cow
cattle operation

Item	Total quantity	Price	Cost or income
			- - U.S.$ - -
Investment			
Land[a]	108 ha	1,480/ha	159,840
Buildings & capital investment[b]			27,027
Equipment[c]			15,262
Livestock[d]			63,750
Total			265,879
Net basic production costs			
Fertilizer[e]	22.5 tons	132/ton	2,970
Nitrogen[f]	3,850 kg	0.58/kg	2,233
Lime[g]	48.9 tons	22/ton	1,076
Hay[h]	150 tons	30/ton	4,500
Range cubes[i]	11,235 kg	0.20/kg	2,247
Grain[j]	9,112 kg	0.17/kg	1,549
Rye & ryegrass pasture[k]	9.7 ha	305.03/ha	2,959
Salt[l]	1,355 kg	0.07/kg	95
Minerals[m]	1,980 kg	0.22/kg	436
Medicine[n]	-	-	599
Veterinary services[o]	6 visits/yr	50/visit	300
Maintenance, fuel, repairs[p]	-	-	4,860
Utilities[q]	-	-	144
Other and administrative	-	-	1,000
Replacement bull	-	-	1,000
Labor, hired	200 hr	4.25/hr	850
Total			26,816
Less salvage income			
Cull cows, 21 at 405 kg	8,505	0.98	8,335
Cull replacement heifers, 3 at 340 kg	1,020	1.11	1,132
Cull bull, 1 at 630 kg	630	1.22	769
Total			10,236
Net basic production costs			16,580

continued

Table 4.1. Costs and returns for example 100-brood-cow
cattle operation (continued)

Item	Total quantity	Price	Cost or income
			- - U.S.$ - -
Other Costs			
Capital costs[r]			
Land			1,200
Buildings and capital investment			3,243
Equipment			1,831
Livestock			7,650
Operating capital			1,609
Subtotal			15,533
Ownership costs[s]			
Depreciation	-	-	4,392
Taxes	-	-	4,043
Insurance	-	-	423
Subtotal	-	-	8,858
Labor, own	760 hr	4.25/hr	3,230
Management	1/7 time	12,000/yr	3,000
Total, other costs			30,621
Total, net basic production and other costs			47,201
Income, calves			
Steers (43 @ 205 kg)	8,815 kg	1.50/kg	13,223
Heifers (17 @ 180 kg)	3,060 kg	1.35/kg	4,131
Total (198 kg)	11,875 kg	1.46 kg	17,354
Net income			
Above net basic production costs			772
Above net basic production and other costs			-29,847
Breakeven or cost of production			
Net basic production costs ($16,582÷11,875 kg)			1.40
All costs ($47,201÷11,875 kg)			3.97

62

Footnotes to Table 4.1. Costs and returns for example 100 brood cow.

---

^aLand
100 ha of Bahia grass and 8 ha of other land or 108 ha total.
^bBuildings and capital investment
(1) Fence, 6.4 km at $930 per km at $5,952.
(2) Feed storage. Galvanized steel building 7m x 9m x 3m high with concrete floor at $4,950. Assume 50 percent of this is charged to the cow/calf operation at $2,475.
(3) Equipment shed, open, with dirt floor, 30m x 11m x 4.5m high at $9,000. Of this, assume 10 percent or $900, is charged to the cattle operation.
(4) Shop, 12m x 9m x 3m high with concrete floor at $8,000. Of this, 10 percent or $800, is charged to the cow/calf operation.
(5) Open pole barn for hay storage, 15m x 11m x 4.5m high at $4,500. Of this, 100 percent or $4,500, is charged to the cow/calf operation.
(6) Four inch well, with all related equipment at $4,000.
(7) Miscellaneous, such as mineral feeders and water tanks at $5,000.
(8) Wooden corrals, with chute at $3,400.
^cEquipment

Item	Cost new	Allocation to cattle operation	Cost to cattle operation
	- U.S.$ -	Percent	- U.S.$ -
70 hp. tractor	17,375	33	5,734
Front end loader	1,100	50	550
Cattle sprayer	975	100	975
Rotary mower	1,825	90	1,643
Pickup truck (1/2 ton)	7,000	40	2,800
Truck (2 ton)	11,000	20	2,200
Fertilizer spreader	6,800	20	1,360
Total			15,262

^dCattle

Item	Number	Value Each	Total
		- - - U.S.$ - - -	
Mature cows	100	500	50,000
Bulls	5	1,000	5,000
Replacement heifers	25	350	8,750
Total			63,750

Footnotes Table 4.1. (continued)

[e]Fertilizer
Average of 225 kg of 10-10-10 per ha annually. Thus, 225x100 ha=2.5 tons.

[f]Nitrogen
110 kg per ha applied on hay land (35 ha) only.

[g]Lime
Hay land limed every 4 years at 2.5 tons per ha. Thus, 35÷4=8.75x2.5=21.9 tons. Other pasture limed every 6 years. Thus, 65÷6=10.8x2.5=27.0 tons. The total is 48.9 tons.

[h]Hay
It is assumed that hay is custom baled and the bales (round) placed in the pastures by the custom baler so virtually no movement is required at feeding time. Cost of harvesting is $30 per ton. Only 141 tons are fed, but 150 tons are harvested. Part of the additional is in square bales and is placed in the hay barn.

[i]Range cubes
See Figure 4.2 and Table 4.2.

[j]Grain
See Figure 4.2 and Table 4.2.

[k]Cost of producing one hectare of rye-ryegrass

Item	Unit	Quantity	Price	Cost/ha
			- - - U.S.$ - - -	
Cash expenses				
Rye seed	kg	85.0	0.42	35.70
Ryegrass seed	kg	22.0	0.66	14.52
Lime[a]	ton	1.0	22.00	22.00
Fertilizer, 5-10-15[a]	kg	570.0	0.13	74.10
Nitrogen[a]	kg	85.0	0.58	49.30
Machinery	--	--	--	15.75
Labor	hr	3.7	3.50	12.95
Land rent, 6 mo.	ha	1.0	75.00	37.50
Subtotal				261.82
Interest[b]	$	261.82	.06	15.71
Total cash expenses				277.53
Fixed costs of machinery				27.50
Total				305.03

[a]Cost spread.
[b]12% for 6 months.

Footnotes Table 4.1. (continued)

[l]Salt

Mature cows	11 kg/cow/yr x 76	=	836 kg
Bulls	11 kg/bull/yr/x 5	=	55 kg
Young heifers	8 kg/heif./yr x 25	=	200 kg
First-calf heifers	11 kg/heif./yr x 24	=	264 kg
Total			1,355 kg

[m]Minerals

Mature cows	16 kg/cow/yr x 76	=	1,216 kg
Bulls	16 kg/bull/yr x 5	=	80 kg
Young heifers	12 kg/heif./yr x 25	=	300 kg
First-calf heifers	16 kg/heif./yr x 24	=	384 kg
Total			1,980 kg

[n]Medicine

Worm medicine, 85 calves @ $1.25 per head	=	$106
Blackleg vaccine, 85 calves @ $0.12 per head	=	10
Growth implants, 50 calves, 2 implants per calf, @ $1.08 per dose	=	108
Fly spray concentrate, 38 liters @ $3.15 per liter	=	120
Insecticide tags, 130 head, 1 tag per cow, heifer and bull, @ $1.25 per tag	=	163
Iodine, 7.6 liters @ $6.60 per gal.	=	50
1,000 ml injectable antibody @ $0.03 per ml	=	30
Uterine boluses	=	12
Total		$599

[o]Veterinary services

Assumes the operation is 50 km from the veterinarian's office.

[p]Maintenance, fuel and repairs

Tractor

Hay feeding, 2 hr per day for 120 days	=	240 hr
Front end loader	=	75 hr
Mowing, 2 times per year	=	83 hr
Fertilizer spreading	=	63 hr
Miscellaneous	=	50 hr
		5 1 1 hr

Var. cost = 511 hr x $3.98/hr = $2,034

Fixed cost = 511 hr x $1.62/hr (4.92/hr x 33%) = $828

Pickup truck: 7,200 km @ $0.15/km = $1,800.

2 ton truck: 12,000 km/yr x 5% to cattle operation = 600 km at $0.55/km = $330.

Rotary mower: Var. cost $2.41/hr x 83 hr = $200, fixed cost $2.79/hr. x 90% = $2.51 x 83 Hr. = $208

Fertilizer spreader: Var. cost $2.00/hr x 63 hr = $126, fixed cost $4.25/hr x 20% = $0.85 x 63 hr = $54

Hired labor is entered under basic production costs. In addition, there is a charge of $3,230 for owner's labor. There is also a charge of $3,000 for management. The total charge for the expenses other than basic production costs is $30,621. Total costs are thus $47,201.

Income from the 205 kg steers, calculated using a price of $1.50 per kg, yields an income of $13,223. Heifer calves, 180 kg average, are assumed to be sold at $1.35 per kg. Total income (not including income from cull animals, which was subtracted from basic production costs to give the net basic production cost) is $17,354. As a result, there would be a net income of $772 above basic production costs, but a loss of $29,847 when all costs are considered.

The breakeven price--that is,the weighted price for both steer and heifer calves--required to break even is $1.40 per kilo if only basic production costs are considered, but $3.97 when all costs are taken into account. The breakeven price, which is also equivalent to cost of production, is calculated by dividing each of the two cost categories by total kilos of calves sold.

Footnotes Table 4.1. (continued)

[q]Utilities
200 kwh/mo at $0.06/kwh = $12/mo or $144/yr
[r]Capital costs
Land, $12 per cow unit
Buildings and capital investment, 12% on investment
Equipment, 12% on investment
Livestock, 12% on investment
Operating capital, 12% on 50% of basic production costs
[s]Ownership costs
Depreciation
  Buildings and capital investment, 5% of $27,027 = $1,351.
  Equipment, 10% of $15,262 = $1,526.
  Cattle, 12% (replacement rate) of $125 per animal average difference between salvage value and replacement cost on 101 head (includes bulls). Thus, 12% of $12,625 = $1,515.
Taxes
  Two percent of value of land, buildings and capital investment and equipment: $202,129 x 2% = $4,043.
Insurance
  One percent of buildings, equipment and other capital investment: $42,289 x 1% = $423.

Table 4.2. Quantities of supplement feed, example cow/calf
operation

Hay		
Mature cows	1.5 tons/cow/yr x 76	114.0 tons
First-calf cows	.5 ton/heif/yr x 24	12.0 tons
Replacement heifers	.3 ton/heif./yr x 25	7.5 tons
Bulls	1.5 tons/bull/yr x 5	7.5 tons
		141.0 tons
Range cubes (30%)		
Mature cows	115 kg/cow/yr x 76	8,740 kg
First-calf cows	80 kg/heif./yr x 24	1,920 kg
Replacement heifers		-
Bulls	115 kg/bull/yr x 5	575 kg
		11,235 kg
Grain		
Replacement heifers	2.7 kg/day/heif.	
	x 135 days x 25	9,112 kg
Rye and ryegrass pasture		
First-calf cows		9.7 ha

Growing

Cattlemen in many parts of the world maintain ownership of calves
from their cow/calf operation beyond weaning or purchase calves with the
intention of deriving benefit from the weight gain. This is called a
growing operation. Another term is stocker operation.

It is assumed in this section that the producer in the last example
decides to grow out the 43 steers from the cow/calf operation and, in
addition, purchase 57 more at the same weight (205 kg). Pasture is
beginning to decline in quality and quantity by the time the growing
operation begins (September), so the calves will be held on essentially a
maintenance ration of hay, grain, and excess bahia grass pasture until the
main forage source, winter rye-ryegrass, is available about December 15.
During the 3 cool-season months from September 15-December 15 the
average total gain per steer is only 23 kg.

The cattle are placed on the cold season rye-ryegrass pasture for
about 135 days, that is, until the end of April or into early May during
which time they gain 98 kg each or 0.72 kg per day. The average gain
per head for the entire 225-day growing period from September through
April is 120 kg or 0.54 kg per day. Steers weigh 325 kg at the end of
the growing period. Total gain for 99 steers sold (1 dies) is 11,675 kg.

The development of a complete enterprise budget (Table 4.3) is
simplified if budgets for the cow/calf operation are already completed
because many expenses in the proposed system, such as buildings and
other capital investment, are based on those budgets. As a consequence,
the only requirement is to allocate the appropriate percentage attributable

Table 4.3.   Example of a growing operation

Item	Total quantity	Price	Cost or income
			- - U.S.$ - -
Investment			
Land			0
Buildings & capital investment			9,500
Equipment			2,270
Livestock			0
Total			11,770
Basic Production costs			
Cattle (100 @ 205 kg each)	20,500 kg	1.50/kg	30,750
Pasture	20 ha	305.03/ha	6,101
Hay	13.5 tons	30/ton	405
Grain	22.5 tons	150/ton	3,375
Salt	350 kg	0.07/kg	25
Minerals	350 kg	0.22/kg	77
Medicine	-	-	213
Veterinary services	-	-	0
Maintenance, fuel, repairs	-	-	500
Labor, hired	-	-	0
Other & administrative	-	-	200
Total, without cattle			14,586
Total, with cattle			41,646
Other costs			
Capital costs			
Land			-
Buildings & capital investment			1,140
Equipment			272
Livestock			0
Operating capital			2,624
Subtotal			4,036
Ownership costs			
Depreciation	-	-	702
Taxes	-	-	235
Insurance	-	-	117
Subtotal	-	-	1,054
Land rent	-	-	0
Labor, own	225 hr	4.25/hr	956

continued

Table 4.3. Example of a growing operation (continued)

Item	Total quantity	Price	Cost or income
		- - U.S.$ - -	
Management	1/14 yr	21,000/yr	1,500
Total, other costs			7,546
Total, basic production & other costs, no cattle			22,132
Total, basic production & other costs, with cattle			49,192
Income, calves (99 @ 325 kg each)	32,175 kg	1.45/kg	46,654
Net income			
Above net basic production costs			5,008
Above all costs			-2,538
Breakeven			
Basic production costs ($41,646÷32,175 kg)			1.29
All costs ($49,192÷32,175 kg)			1.53
Cost of gain			
Basic production costs ($14,586÷11,675 kg)			1.25
All costs ($22,132÷11,675 kg)			1.90

to growing. This allocation process would be similar to the one done for the cow/calf operation.

One difference between a cow/calf and growing operation is that in the latter system cattle are a production cost rather than an investment because they are all purchased and sold in the production year. But costs other than basic production costs are calculated in a manner similar to that for the cow/calf operation. Interest or opportunity cost is only calculated on a portion of operating capital (90 percent in this example) because money is not tied up during the entire 225 day period. The percentage is relatively high because a major part of the cost is steer purchase (or transfer) at the beginning of the production period.

The 99 steers are sold at $1.45 per kg, which is considerably above the $1.29 required to break even on basic production costs but somewhat below the $1.53 per kg needed to cover all costs. The cost of gain, calculated by dividing the "without cattle" costs by the total gain, is $1.25 per kg for basic production costs only and $1.90 if all costs are included.

The budget just presented is the type analysis that would be prepared by a person considering growing out cattle. The careful manager would also prepare one of these budgets as a "closeout" sheet to determine profit or loss and for comparison with initial estimates to sharpen the next year's estimates. In addition, a farmer and a cattle owner who are considering entering into an agreement for the farmer to grow out cattle on contract could use this information to set the charge.

Semiconfinement Grain-Finishing Operation

The continuation of enterprise budgeting in this section assumes that the 99 grown-out cattle, which come off winter pasture in late April or early May, are grain fed to finished weight (461 kg) on the farm rather than being sold as feeders. They are thus assumed to be transferred to the feeding operation at $1.45 per kg weighing 325 kg each, the same assumed sale price and weight as the sale of grown-out cattle.

The steers are held in a strongly fenced one ha pasture and fed in open bunks for 120 days. There is a net gain of 135 kg or an average daily gain of 1.125 kg based on the net market weight after shrink, and a 0.50 percent death loss (1 animal) is subtracted from the gross weight.

Total investment is estimated at $11,704. Cattle, just as in the growing operation, are shown as a production cost rather than as an investment (Table 4.4). Basic production costs without cattle are $23,469 while cost are $70,123 with cattle. Other costs, such as labor, management, capital and ownership costs add another $11,911.

Total income, assuming a selling price of $1.54/kg, is $69,574. There is a forward or positive margin of $0.09/kg because the purchase (or transfer price) is $1.45. It is very important that these transfer costs be carefully documented and allocated to each enterprise so that an assessment can be made of which enterprise is making or losing money.

There is a loss of $549 above basic production costs as well as a loss of $12,460 above all costs from the grain-finishing phase. The breakeven price (calculated by dividing cattle production cost by total kilos sold) is $1.55 per kg. This is a useful figure to calculate prior to feeding the cattle, or to estimate while they are on feed, as opportunities will often arise to fix a sale price by forward contracting, (making a contract for sale at a specified price and date).

The cost of gain is calculated by dividing total costs by total net gain. In this case, cost of gain lies somewhere between $1.80 and $2.72 per kg depending on the items included in the cost estimates. The exposition in Table 4.4 demonstrates that care must be taken in interpreting rates of gain and production cost.

Forage Finishing

A farm plan (only one among many potential ones) for finishing steers on forage and also one that combines short-term finishing with forage (the next alternative evaluated) is given in Figure 4.3 on a chronological basis. The overlaps shown in the figure are to allow time for seeding and growth. As can be appreciated, management is critical because each forage must be ready at a certain time.

70

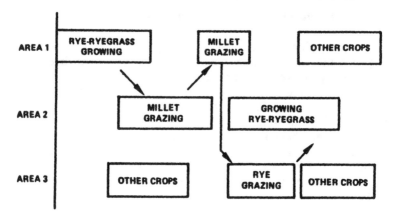

FORAGE ONLY, GROWING AND FINISHING TO SLAUGHTER WEIGHT

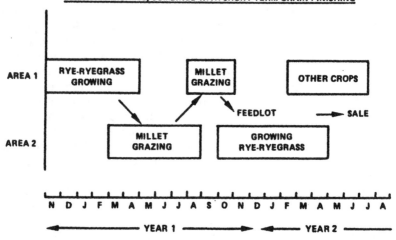

FORAGE FINISHING, COMBINED WITH SHORT-TERM GRAIN FINISHING

FIGURE 4.3   EXAMPLE FARM PLAN FOR FORAGE FINISHING
STEERS, AND FORAGE FINISHING COMBINED WITH
SHORT-TERM GRAIN FINISHING.

Table 4.4. Example-grain finishing operation

Item	Total quantity	Price	Cost or income
			- - U.S.$ - -
Investment			
Land	1 ha	1,480/ha	1,480
Buildings & capital investment			9,000
Equipment			1,224
Livestock			0
Total			11,704
Production costs			
Cattle (99 @ 325 kg each)	32,175 kg	1.45/kg	46,654
Pasture expenses			0
Hay	27.2 tons	50/ton	1,360
Grain mix	118,800 kg	0.18/kg	21,384
Salt	89 kg	0.07/kg	6
Minerals			0
Medicine	-	-	388
Veterinary services	1 visit	50/visit	50
Maintenance, fuel, repairs	-	-	106
Labor, hired	-	-	0
Other and administrative	-	-	175
Total, without cattle			23,469
Total, with cattle)			70,123
Other costs			
Capital costs			
Land			178
Building & capital investment			1,080
Equipment			147
Livestock			0
Operating capital			7,573
Subtotal			8,978
Ownership costs			
Depreciation	-	-	572
Taxes	-	-	234
Insurance	-	-	117
Subtotal	-	-	923
Land rent	-	-	0
Labor, own	120 hr	4.25/hr	510

continued

Table 4.4. Example-grain finishing operation (continued)

Item	Total quantity	Price	Cost or income
		- - U.S.$ - -	
Management (1/14 time)	120 hr	21,000/yr	1,500
Total, other costs			11,911
Total, production & other costs, no cattle			35,380
Total, basic production & other costs, with cattle			82,034
Income, (98 steers @ 461 kg)	45,178 kg	1.54/kg	69,574
Net income			
Above basic production costs			-549
Above all costs			-12,460
Breakeven			
Basic production costs ($70,123÷45,178 kg)			1.55
All costs ($82,034÷45,178 kg)			1.82
Cost of gain			
Basic production costs ($23,469÷13,003 kg)			1.80
All costs ($35,380÷13,003 kg)			2.72

The example farm plan for a combination growing and forage-only finishing operation is quantitatively presented in Table 4.5. Cattle come off the rye-ryegrass growing pasture about May 1 and go on millet that was planted about the middle of March. They remain on this stand for 120 days, at which time they are moved to another 14 hectare millet field for an additional 65 days. Then, about the first week of November, the cattle are moved to bahia pasture (or corn stalks and grain) as it is too late for warm-season forages. The cattle, which by this time weigh between 410 and 425 kg, are maintained on this feeding program for 35 days during which time they receive 2.7 kg of grain mixture per day per head. Then the cattle are moved to rye pasture for the last 70 days of the finishing period and finally sold at a payweight (weight after shrink and marketing costs are deducted) of 465 kg.

The rye pasture used in the final finishing phase could last beyond February 10, but the cattle must be moved so the land can be prepared for other crops such as corn. Failure to move the cattle at this time means giving up a large probable return from farming for a relatively small additional return from cattle. No cost of land use is charged on the rye because it is used during a period of the year when row crops are not grown. In contrast, an opportunity land-use charge of $37.00 per ha is assessed to the forages when row crops could have been grown.

Table 4.5. Farm plan for growing followed by forage only finishing of steers to slaughter weight

Area no.	Use and forage type	Total land required	Carrying capacity	Dates	Total time on forage	Per head			
						Weight		Total gain	Gain per day
		- Ha -	- Hd/ha -		-Days-	In	Out	- - Kg - -	- - - - -
1	Growing, Rye-ryegrass	20	5	Dec. 15 May 20	165	205	325	120	0.73
2	Grazing No. 1 Millet	10	10	May 1- July 31	120	325	380	55	0.46
1	Grazing No. 2 Millet	14	7	Aug. 1- Nov. 5	65	380	410	30	0.46
-	Bahia pasture, or corn stalks & grain	-	-	Nov. 6 Dec. 1	35	410	425	15	0.43
3	Rye	27	3.67	Dec. 1- Feb. 10	70	425	465	40	0.57

The cost analysis format (Table 4.6) is similar to the growing operation. The major production expense, apart from cattle, is forages, which constitute about 85 percent of the total. Interest on operating capital ($5,381) is another major expense, which is computed on 90 percent of the cattle production costs because most of the money is tied up from the beginning of the operation. However, the expense is only charged for 9 months as cattle are only fed for 290 days.

Beef from cattle that are completely finished on forage generally have a yellow colored fat and consequently may not receive the premium that would be derived from grain finished beef, at least in some countries. The meat is usually less marbled (leaner). A price of $1.21 per kg, roughly equivalent to that of cull bulls , thus is used for this example. The result is a loss of $11,292 above basic production costs and $22,599 above all costs. A price of at least $1.46 per kilo would be required to cover minimum production costs and $1.71 for all costs.

It is significant that although the cost of gain ($1.48 on basic production costs) is much lower than for grain finishing ($1.80), there is also a much larger total loss on the operation. This data, as well as the

Table 4.6. Forage finishing example

Item	Total quantity	Price	Cost or income
		- - U.S.$ - -	
Investment			
Land			0
Buildings & capital investment			10,540
Equipment			1,574
Livestock			0
Total			12,114
Basic production costs			
Cattle			
(99 steers @ 325 kg each)	32,175 kg	1.45/kg	46,654
Forage expenses			16,043
Hay			0
Grain	9,355 kg	0.18/kg	1,684
Salt	891 kg	0.07/kg	62
Minerals	891 kg	0.22/kg	196
Medicine	-	-	438
Veterinary services	1 visit	50/visit	50
Maintenance, fuel, repairs	-	-	322
Labor, other	190/hr	4.25/hr	808
Other and administrative	-	-	175
Total, without cattle			19,778
Total, with cattle			66,432

continued

Table 4.6. Forage finishing example (continued)

Item	Total quantity	Price	Cost or income
			- - U.S.$ - -
Other costs			
Capital costs			
Land			0
Buildings & capital investment			1,265
Equipment			189
Livestock			0
Operating capital			<u>5,381</u>
Subtotal			6,835
Ownership costs			
Depreciation	-	-	684
Taxes	-	-	242
Insurance	-	-	<u>121</u>
Subtotal	-	-	1,047
Land rent	-	-	888
Labor, own	100 hr	4.25/hr	425
Management	2/14 time	21,000/yr	3,000
Total, other costs			11,307
Total, basic production & other costs, no cattle			31,085
Total, basic production & other costs, with cattle			77,739
Income, (98 steers @ 465 kg)	45,570 kg	1.21/kg	55,140
Net income			
Above basic production costs			-11,292
Above all costs			-22,599
Breakeven			
Basic production costs ($66,432÷57,570 kg)			1.46
All costs ($77,739÷45,570 kg)			2.32
Cost of gain			
Basic production costs ($19,778÷13,395 kg)			1.48
All costs ($31,085÷13,395 kg)			2.32

cost of gain calculations, show how important it is to carefully specify both costs and sale price in the budget.

Combination Forage and Short-Term Grain Finishing

This section provides an analysis in which cattle from the growing operation are fed on the two stands of millet and then, rather than being placed on ryegrass, are finished out on grain. During this final phase, a grain mixture consisting mainly of shell corn is fed to the steers for 60 days, during which time they achieve an average daily gain of 1.09 kg per day. The total finishing operation covers 245 days, during which time the cattle gain a net 150 kg for a final sale weight of 475 kg.

Production costs (Table 4.7) not including cattle are $20,245, which is about $450 more than the forage-only finishing approach. But final weight is 10 kilos more per animal, and the price received is $1.50 per kg rather than $1.21 per kg because the fat is now white in color and a premium price is paid.

In contrast to the loss suffered from forage-only finishing, this analysis indicates a profit of $2,926 above basic production costs, even though indirect or opportunity costs are not completely recovered.

Table 4.7. Combination example-forage and grain-finishing operation

Item	Total quantity	Price	Cost or income
		- - U.S.$ - -	
Investment			
Land	1 ha	1,480/ha	1,480
Buildings & capital investment			11,030
Equipment			1,574
Livestock			0
Total			14,084
Production costs			
Cattle			
(99 steers @ 325 kg each)	32,175 kg	1.45/kg	46,654
Forage expenses			7,941
Hay			0
Grain	59,400 kg	0.18/kg	10,692
Salt	396 kg	0.07/kg	28
Minerals	396 kg	0.22/kg	87
Medicine	-	-	388
Veterinary services	1 visit	50/visit	50
Maintenance, fuel, repairs	-	-	268
Labor, hired	145/hr	4.25/hr	616
Other and administrative	-	-	175
Total, without cattle			20,245
Total, with cattle			66,899

continued

Table 4.7. Combination example forage and grain finishing
operation (continued)

Item	Total quantity	Price	Cost or income
			- - U.S.$ - -
Other costs			
Capital costs			
Land			178
Buildings & capital investment			1,324
Equipment			189
Livestock			0
Operating capital			4,817
Subtotal			6,508
Ownership costs			
Depreciation	-	-	709
Taxes	-	-	282
Insurance	-	-	141
Subtotal	-	-	1,132
Land rent	-	-	888
Labor own	100 hr	4.25/hr	425
Management	(1/14 time)	21,000/yr	1,500
Total, other costs			9,321
Total, production & other costs, no cattle			29,566
Total, production & other costs, with cattle			76,220
Income, (98 steers @ 475 kg)46,550 kg		1.50/kg	69,825
Net income			
Above basic production costs			2,926
Above all costs			-6,395
Breakeven			
Basic production costs ($66,899÷46,550 kg)			1.44
All costs (76,220÷46,550 kg)			1.64
Cost of gain			
Basic production costs (20,245÷14,375 kg)			1.41
All costs (29,566÷14,375 kg)			2.06

Integrated System

This last section evaluates a combination operation from birth to sale at slaughter weight. The approach is slightly more involved than just adding up net income figures as some items would be duplicated that way. A summary of this approach, given in Table 4.8, shows that $7,121 would be earned above basic production costs while a loss of $40,367 would be incurred when all costs are considered. Care must be taken in preparation of this table to subtract income from heifer calves as well as income from cull animals to arrive at net basic production costs. The steers would be 22.7 months of age (9 to weaning, 5.5 for backgrounding, 8.2 for finishing) when finally sold.

The complete production phase spans 25.7 months (including 12 months in the cow/calf phase of which about 4 months are postweaning). Net income per day to the producer is thus $9.19 per day, which can be interpreted as a return to factors of production, that person's labor and management, and ownership costs. Society as a whole benefits by reduced death loss, shrinkage and so forth because cattle do not have to be moved from one area to another.

A summary table of economic relationships for the various systems is provided in Table 4.9. The best option depends on each producer's situation and, of course, on the assumptions about prices. Total net income is highest for the completely integrated operation, but this requires two years so that, on a per day basis, the growing only option yields the highest return. The combination forage/grain-finishing system also appears to be viable, but there are many possible drawbacks, such as variability in weather conditions, that could result in a poor stand of millet and a need for very good management.

The analysis shows that income is not sufficient in any system to cover total costs. Because many of these costs, such as operating capital, are tied up in cattle, then labor, management and depreciation are essentially opportunity losses by the owner--that is, there is no real cash outflow. Nevertheless, these costs should be calculated so that realistic financial comparisons can be made, especially regarding fixed facility replacement.

The importance of preparing careful footnotes has also been brought out for it is only by writing down the various assumptions that changes in the budgets can be made quickly and easily. Another point made is that care should be taken when comparing breakeven and gain costs with those prepared by other individuals as there will be considerable differences in items included or left out of calculations. Perhaps the major point has been that although budgeting may appear to be relatively simple, in reality it is rather complex. Only after this basic, powerful tool is mastered can a wide variety of more sophisticated techniques such as linear programming can be employed.

Table 4.8. Example integrated system, cow/calf to finishing of steers

	Basic production costs[a]	Minus		Plus		Net costs	
		Salvage income	Heifer calves	Steers purchased[b]	Other costs	Basic production	All
	- - - - - - - - - - - - - - - - - - U.S.$ - - - - - - - - - - - - - - - - -						
Costs							
Cow/calf, usual weaning	43,070	26,816	10,236	4,131	-	30,621	12,449
Growing	37,556	14,586	-	-	15,424	7,546	30,010
Combination forage & grain finishing	20,245	-	-	-	9,321	20,245	29,566
Total	110,192						62,704
Income (98 steers at 475 kg ea)						69,825	69,825
Net Income						7,121	-40,367
Breakeven[c]						1.35	2.37
Net income per day[d]						9.19	-

[a] Without cattle
[b] 100 minus 43 from cow/calf operation = 57 purchased, $1.32/kg, 205 kg each
[c] Cost divided by kilos sold (46,550)
[d] 775 days (365 cow/calf, 165 growing, 245 days combination forage and grain finishing)

Table 4.9. Summary of example systems

Item	Units	Cow/calf	Growing	Grain only finishing	Forage only finishing	Combination forage and grain finishing	Integrated cow/calf to finishing
Basic production							
costs	$	16,580	41,646	70,123	66,432	66,899	62,704
All costs	$	47,201	49,192	82,034	77,739	76,220	128,173
Net income above							
Basic production							
costs	$	772	5,008	-549	-11,292	2,926	7,121
All costs	$	-29,847	-2,538	-12,460	-22,599	-6,395	-58,348
Marketing							
Weight	kg	198	325	4.61	465	475	475
Price/kg	$	1.46	1.45	1.54	1.21	1.50	1.50
Breakeven above							
Basic production							
costs	$	1.40	1.29	1.55	1.46	1.44	1.35
All costs	$	3.97	1.53	1.82	1.71	1.64	2.75
Cost of gain							
Basic production							
costs	$	-	1.25	1.80	1.48	1.41	-
All costs	$	-	1.90	2.72	2.32	2.06	-
Return per day, above							
Basic production	$	2.12	30.35	-4.58	-38.94	11.94	9.19

CHAPTER 5

PARTIAL BUDGETING:
TECHNIQUES AND APPLICATION

The last chapter was oriented toward whole-farm planning and complete enterprise budgets. There is one technique, partial budgeting, that is an intermediate between the two and is a type of marginal analysis that is well suited to evaluate small changes in livestock systems. This technique is now explained and a number of examples provided.

The Partial Budgeting Technique

Partial budgeting is used to evaluate the economic impact from a change in business operating procedure. Partial budgeting differs from an enterprise budget in that several enterprises might be involved in the change, but a partial budget is not appropriate to prepare a whole-farm plan (Kay, 1981). Partial budgeting only uses the data that is absolutely necessary for the evaluation process. Some diverse examples where this technique is appropriate include comparison of whether new inputs or techniques will improve net income. For example, is it less expensive to hire someone to move livestock to water or to develop water facilities, or whether forage A or forage B is the least-cost alternative.

A partial budget can be defined as a systematic listing of the possible changes in costs and returns in a given time period when production practices change. The analyst may elect to use categories from complete enterprise budgets or simply develop the budgets on an ad hoc basis. If the purpose of the analysis is to analyze an entire system, such as the introduction of grass finishing steers as a supplement to a cow/calf operation, it is likely that a complete enterprise budget should first be developed because the objective is to determine the impact on the whole system. This is especially true, as shown in the last chapter, where there are significant interactions between production units. On the other hand, evaluation of whether to buy brand X or brand Y of an electrical generator would only require the relevant costs, providing the benefit is the same with either one.

Partial budgets are based upon four possible kinds of changes:

1. Added income resulting from the proposed changes; this revenue source includes increased value from existing outputs as well as new income sources
2. Reduced income from either reduction or elimination of certain income sources
3. Added costs as a result of the new or expanded undertaking
4. Reduced costs because of enterprises or practices that would be reduced or eliminated

The net change in income (revenue) from the proposed investment is simply the difference between the gains (added income plus reduced costs) and the losses (reduced income plus added costs). Partial budgeting does not provide an estimate of the net income that can be gained from an overall operation, although the changes can be introduced into complete budgets to carry out such computations. The approach of using complete budgets as a base is taken in this chapter's first example, which is an extension of the cow/calf budget set forth in the previous chapter.

Early Weaning: An Example

One problem producers face is to determine the optimal time to sell their various classes of livestock. Usually there are seasonal price fluctuations that come about due to a variety of reasons. For instance, much of the tropical world has a regular dry season during which cattle and small stock may lose weight. The tendency is for owners to increase sales just before or early in that period in order to adjust herd numbers to reduced forage conditions. As a result of expanded supply, price normally will drop. Consequently, one alternative is for a producer to wean and sell calves a few months prior to the seasonal price decline.

As an example, recall the producer who had 100 cows and weaned 43 steer calves and 17 heifer calves at an average of 205 kg and 180 kg respectively. Selling the calves at lighter than normal weight by weaning early has several advantages for this particular situation:

1. Lightweight calves sell at a higher price per kg than do heavier calves (in this example it is determined that the average price for 150-225 kg calves is 8 percent higher in June than in September)
2. The income can be reinvested to earn interest, or early payments can be made on loans
3. Production expenses are reduced
4. Higher calf crops might be expected in the longer term as there is less stress on the cow
5. The operator can use management skills in other enterprises
6. Price risk is reduced

The approach to analyze this alternative is to prepare an abbreviated budget similar to the basic cow/calf one (Table 4.1) and to compare the results. However, only items that change or are required for computational purposes need be included. The estimated costs and returns for the early weaning alternative are presented; one column shows the difference between early weaning and usual weaning (Table 5.1). The analysis indicates that if calves are weaned 60 days early, two major production costs, fertilizer and nitrogen, could be reduced by $772, thereby resulting in a 3 percent reduction in basic production costs. Also, operating capital requirements would be reduced along with labor and management time. These savings in the "other costs" category are estimated at $820.

The steer calves are assumed to be sold at 170 kg (rather than 205 kg) and the heifers at 145 kg (rather than 180 kg). The lighter weight steer calves are assumed to sell at a premium of $0.20 per kg while the heifer calves bring $0.15 per kg more. That premium includes seasonality, that is, the fluctuations within a year. The net result under these assumptions is that the early weaning alternative produces a net revenue of $455 less than the usual weaning alternative above net basic production costs, but a benefit of $363 when all costs are considered. In effect, the two alternatives are about similar. The operator would want to watch price movements to take advantage of favorable changes.

The main point demonstrated in this section is the simplicity with which management alternatives can be evaluated by using partial budgeting, once a basic budget has been prepared. Other production practices can also be readily evaluated, and attention is now turned to one of them.

Preconditioning Calves for Sale to Feedlots

As another example problem suppose that a cattle producer wants to precondition calves from his or her operation rather than selling them immediately after weaning. "Preconditioning" is a calf management program in which calves are provided with medication and taught to eat from feed bunks prior to shipping to feedlots. The objective is to reduce the incidence of disease, death loss and poor performance of weaned calves during the first few weeks of feeding (Pate and Crockett, 1979). Preconditioning programs are largely an outgrowth of concern from national-level planners about ways to make the whole beef system more efficient and from cattle feeders interested in reducing their own costs. But it is the primary calf producer who must pay the cost with the expectation that economic rewards will accrue from a premium paid for the calves as well as for any additional weight gain. The feeding of livestock to improve their physical condition prior to trekking or moving them long distances to market is a type of preconditioning program.

Assume that the preconditioning feed is formulated with ingredients typically found in a supplement ration as well as urea to adapt the rumen of the calves to this ingredient. The supplement is initially fed at 0.09 kg per head per day and increased during a 10 to 14-day period until calves are consuming an average of 4.6 kg per day. A perennial grass

Table 5.1. Example of weaning alternative

Item	Total quantity	Price	Cost or income early weaning	Change from usual weaning time
			- - - - - - - $US - - - - - - -	
Basic production cost				
Fertilizer[a]			2,640	-330
Nitrogen[b]			1,791	-442
Other			21,613	0
Total[c]			26,044	-772
Salvage income			10,236	0
Net basic production costs			15,808	-772
Other costs				
Capital costs[d]				
Operating capital			1,287	-322
Other			13,924	0
Ownership costs			8,858	0
Labor[e]			2,972	-258
Management[f]			2,760	-240
Total other costs			29,801	-820
Total net basic production and other costs			45,609	1,592
Income, calves				
Steers				
(43 at 170 kg)	7,310/kg	1.70/kg	12,427	796
Heifers				
(17 at 145 kg)	2,465/kg	1.50/kg	3,698	-433
Subtotal	9,775/kg	1.65/kg	16,125	-1,229
Net income				
Above net basic production cost			317	-455
Above net basic production and other costs			-29,484	363

[a]Fertilizer. Reduce by 10 percent as less is required in the warm rainy season due to reduced grazing from both calves and cows, which require feed only for maintenance.
[b]Nitrogen. It is estimated that nitrogen can be reduced by 10 percent because less hay is required as cows are in better condition going into the winter.
[c]Hay. It is estimated the same amount of hay will be harvested even though less nitrogen is used because the cows can be kept on less land in an early weaning program. This will allow more land for hay production.

greenchop is offered <u>ad libitum</u> throughout the preconditioning period. The preconditioning pens, which have 19 m^2 of space per calf, have an earthen floor, wooden feed bunks, open water troughs, and are partially covered to protect the feed bunks from rain. Calves are given medication as noted in Table 5.2.

The estimated cost for the 25-day preconditioning program in accordance with the input/output price ratios used for examples in the last chapter is $35.90. An additional gain of 11 kilos per calf is attributed to the program. Another benefit is a $0.03 per kilo premium so that a gross additional return of $17.19 is obtained. The net result of this partial analysis is a projected loss of $18.71 per head. Calculations of the breakeven selling price, $1.07, indicate that price would have to move upward substantially to make the program pay.

The format is slightly different in this table than the one described in the theoretical part of this chapter and in the previous example. This is because there is no change in costs or income in the traditional system. Other examples in which this same analytical framework might be applied is partial fattening of weaned feeder pigs and raising dairy calves for sale as vealers. The basic concept is that a person has an animal and wants to know if the additional cost of improving its condition before sale is justified. Let us take another related example.

Supplemental Feeding of Steers on Pasture

The example situation used in Chapter 3 to explain the "how much to produce" production economics problem is now expanded. Suppose that the farm operator uses all land for a pasture-only fattening operation in which steers are traditionally purchased at 375 kg and sold for slaughter at 438 kg. The problem is to determine whether addition of supplemental concentrate feed will increase net income.

---

[d]Capital costs. Cattle are sold 2 months earlier under the early weaning option, and thus less operating capital is required. Assume it is reduced by 20 percent.

[e]Labor. Calves and cull animals are sold 2 months early. Even though not quite 2 months of labor are saved as there is still breeding herd work required, it is assumed an equivalent of 2 months of labor is saved--that is, there is an 8 percent reduction in labor costs.

[f]Management. It is assumed that the management savings is equivalent to 1 month. Normal weaning has a management equivalent of $3,000 devoted to the cattle operation. An 8 percent reduction would result in a charge of $2,760.

Table 5.2. Example costs and returns per head from preconditioning calves for 25 days

Item	Quantity per head	Price per unit	Cost per head
	- - - U.S. Dollars - - -		
Additional Costs			
Feed			
Mixed ration	93 kg	0.28/kg	26.04
Greenchop (dry matter basis)	23 kg	0.04/kg	0.92
Medication (vaccines, wormer, grub control)			1.50
Labor (additional in working calves for medication)			0.50
Feeding, facilities, and labor		0.15/hd/day	3.75
Interest ($174 value/calf @ 12% per annum)			1.45
Death loss ($174 value/calf @ 1%)			1.74
Total			35.90
Additional income			
Value at weaning	185 kg	0.94	173.90
Value after preconditioning	196 kg	0.97	191.09
Gross additional return			17.19
Net return to preconditioning			-18.71
Breakeven selling price[a]			1.07

a $\dfrac{\text{Calf value at weaning (\$U.S.173.90)} + \text{Preconditioning cost (\$U.S. 35.90)}}{\text{Selling weight of preconditioned calf (196 kg)}}$ = Breakeven

Information on inventory, number of animals marketed annually, and production measures is provided in Table 5.3 for the basic steer operation without supplemental feeding and also for the proposed operation with supplemental feeding.

No Supplemental Feeding

Now that the basic physical and economic information has been assembled the analytical phase can begin. A budget is prepared that is divided into the various parts outlined in the last chapter: investment, basic production costs, other costs, totals and per unit measures. Total

Table 5.3. Inventory, production measures and economic
relationships for 100 AU steer example operation

Items	Units	Basic no supplement	Proposed with supplement
Inventory			
Total animal units	AU	101	101
Pasture land	ha	100	100
2-year-old steers	no	100	100
Horses	no	1	1
Total animals	no	101	101
Marketing			
Number steers marketed	no	99	99
Production measures			
Death loss	pct	1	1
Gain per day	kg	0.30	0.80
Length of feeding	days	210	210
Purchase weight	kg	375	280
Gain per head	kg	63	168
Sale weight	kg	438	448
Total sale weight	kg	43,262	44,352
Total gain	kg	5,762	12,544
Economic information			
Purchase price, per kg	$U.S.	0.80	0.80
Sale price, per kg	$U.S.	0.85	0.87
Supplement cost, per kg	$U.S.	-	0.26

investment for the no-supplement alternative is $52,100 (Table 5.4).
There is no entry in the cell for livestock because steers are an operating expense rather than an investment item. Total basic production costs
amount to $32,285, of which $30,000 is for cattle. Other costs total
$4,004, of which operating capital constitutes 54 percent. There is an
annual net income above basic production costs of $4,573 and above all
costs of $569.

With Supplemental Feeding

The proposed change to include supplement feed is analyzed using
the input optimization principle of MIC = MVP. The beef production
relation with various levels of feed use is provided in Table 5.5 along
with the MPP, MVP and MIC. Given a feed cost of $0.26 per kg, the
optimal input use is 1.0 kg per day per head. Total feed for the 210-day
feeding period is 21,000 kg or a total cost of $5,460. There are a few

Table 5.4.   Costs and returns for example 100 AU steer
operation with and without supplemental feed

Item	Basic, no supplement	Proposed with supplement
	- - - - - U.S.$ - - - -	
Investment		
Land	50,000	50,000
Buildings and capital investment	2,000	2,000
Equipment	100	500
Livestock	-	-
Total	52,100	52,600
Basic production costs		
Purchased steers[a]	30,000	22,400
Supplemental feed[b]	-	5,460
Labor, hired	-	300
Salt and minerals	30	30
Repairs and maintenance		
Buildings and improvements	50	75
Machinery and equipment	50	75
Veterinary services and supplies	350	350
Seed and fertilizer	1,500	1,500
Machinery		
Operating	255	255
Hired	-	-
Transportation	-	-
Utilities	-	-
Other and administrative	50	5
Total	32,285	30,495
Other costs		
Capital costs[c]		
Land	500	500
Buildings and capital investment	240	240
Equipment	12	12
Livestock	-	-
Operating capital	<u>2,147</u>	<u>2,028</u>
Subtotal	2,899	2,780

continued

Table 5.4.   Costs and returns for example 100 AU steer
operation with and without supplemental feed
(continued)

Item	Basic no supplement	Proposed with supplement
	- - - - - U.S.$ - - - - -	
Ownership costs[d]		
Depreciation[d]	105	105
Taxes	50	50
Insurance	-	-
Subtotal	155	155
Labor, own	450	450
Management	500	600
Total, other costs	4,004	3,985
Total, basic production & other costs	36,289	34,480
Income	36,858	38,586
Net income		
Above basic production costs	4,573	8,091
Above all costs	569	4,106

[a]100 steers at 375 kg each = 37,500 kg times $0.80 = $30,000

[b]Supplement fed at 1.0 kg per head per day for 210 days = 210 kg per head times 100 head = 21,000 kg at $0.26 per kg = $5,460

[c]Capital costs

Land
$5 per AU
Buildings, capital investment, and equipment
12 percent of investment
Operating capital
12 percent annual rate on 95 percent of basic production costs for 210 days (7 months) for no supplement feeding. Same rate, but on 95 percent of basic production costs for 7 months with supplemental feeding

[d]Ownership costs

Depreciation
5 percent of buildings, capital investment and equipment

Table 5.5. Example of determination of optimal supplement
           feed level for steers

Daily ration per head	Daily beef production	Marginal physical product MPP	Marginal value product MVP	Marginal input cost MIC
- - - - - - - Kilos - - - - - - -			- - - U.S.$ - - -	
0.00	0.30			
		0.10	0.09	0.26
0.50	0.35			
		0.30	0.26	0.26
1.00	0.50			
- -				
1.50	0.60			
		0.20	0.17	0.26
2.00	0.70			
		0.20	0.17	0.26
2.50	0.80			
		0.20	0.17	0.26
3.00	1.88			
		0.16	0.14	0.26
3.50	1.93			
		0.10	0.09	0.26
4.00	1.99			
		0.12	0.10	0.26
4.50	1.04			
		0.10	0.09	0.26
5.00	1.08			
		0.08	0.70	0.26

MPP = $\dfrac{\text{Change daily beef production}}{\text{Change supplement}}$

MVP = MPP times output price ($0.85 per kg)

MIC = Price of supplement ($0.26 per kg)

other changes in the budget (Table 5.4) with the result that net income is
$8,091 above basic production costs or double the no-supplement feed
strategy. Net income above all costs is about 7 times greater than the
traditional approach.

Determination of whether to feed supplement, and the optimal
amount, could be carried out by simply using marginal analysis. But the
budget is required to calculate total economic impact. Only a few
numbers actually change in the budget, but the entire budget is reproduced to ensure that mathematical errors are not made.

Parasite Control

Animal health analysis can be carried out in a variety of ways, but two of the most important are modeling and partial budgeting. The latter method is especially important to determine if a practice should be recommended and to determine the extent of benefits. Partial budgeting is a very powerful tool that can provide the relatively quick answers often demanded by veterinarians or planners. An example from Florida is now given on calculations of benefits from parasite control.

Cattle parasites are divided into two classifications, internal and external. Examples of internal parasites now presented are liver flukes, intestinal worms, and grubs. The external ones are horn flies and lice. A myriad of others that plague developing countries, such as tick born diseases, could be evaluated. However, sufficient examples are now given to provide the method for this type of analysis.

A summary of the way in which treatments have been determined for the five parasites is provided in Table 5.6. A table of this type is the first step as it provides a complete description of what has to be done and the cost.

Multiplication of the cost per treatment times the recommended number of treatments for each type of parasite yields the annual cost for each of the three classes of cattle analyzed: mature cows and bulls, replacement heifers, and calves. Recommended treatments and associated costs for intestinal worms vary with animals on a high, medium and low plane of nutrition. Three control methods for horn flies are evaluated. Costs include a charge for additional labor (when applicable) as well as materials and equipment.

Benefits from parasite control are very difficult to quantify due to a multitude of physiological interactions and various geographical/climatic conditions. Consequently, well-documented data and the benefits from control of the various parasites are seldom available. But this problem can be overcome by using estimates from production and veterinary specialists. The estimates can then be divided into two levels, low and high, with the expectation being that the actual benefits will fall within those limits. In all cases, efforts should be conservative in the estimation procedure.

Potential benefits in the example provided are derived from a reduction in death loss, a reduction in weight loss, increased calf crop, or a combination of them. It is estimated, for example, that mature cow death loss in areas infested with liver flukes can be reduced by 2 to 4 percent with the recommended treatments (Table 5.7). Another way of saying this is that liver flukes account for about 2-4 percent death loss in mature cows and treatment is assumed to stop these losses. Flukes are also estimated to cause cows to lose about 18-45 kg, a loss that is realized when they are sold as culls. In addition, mature cows with liver flukes probably have a 6-12 percent reduction in calf crop if they are not treated.

An economic benefit, in dollars per cow unit (one cow and one calf, that is, 1 AU), is calculated for each of the three benefit categories by

Table 5.6. Cattle parasite treatment determination, recommended treatments, cost per treatment, and annual cost per cow unit, Florida, USA

Type of Parasite	Treatment Determination	Recommended Treatment		
		Mature	Replacement	Calves
Internal				
Liver flukes	Fluke-only areas	1/yr	1/yr	-
Intestinal				
worms[a]	High nutrition	0/yr	1/yr	0
	Medium nutrition	1/yr	2/yr	0
	Low nutrition	2/yr	2/yr	1
Grubs	All Florida	1/yr	1/yr	-
External				
Horn flies[b]				
Spray[c]	50 or more flies/animal	6/yr	6/yr	-
Dust bags[d]	50 or more flies/animal	8 mo./yr	8 mo./yr	-
Ear tags	50 or more flies/animal	2/an/yr	2/an/yr	-
Lice[e]	Presence of lice	2/yr	2/yr	-

continued

multiplying the various production loss estimates by the appropriate values. For example, multiplying the high level in death loss (4 percent) from flukes for mature cows times the projected value of $333.00 per cow (a 432 kg cow at $0.77 per kg) gives $13.32 as the annual loss (Table 5.8). Weight loss is calculated by multiplying the weight loss times value per kilo times culling rate, which is assumed to be 10 percent for cows and 3 percent for heifers. The high estimation of weight loss per cow of 45 kg times $0.77 per kg = $35.00 per cow which times 10 percent is $3.50.

Reproductive loss is calculated by multiplying the average value per calf of $235.00 (188 kg calf at $1.25 per kg) times the reproductive loss in percent, adjusted for a herd composition of 90 percent for mature cows and 10 percent replacements. As an example, the high estimate of losses from flukes in mature cows is 12 percent times $235.00 = $28.20. That coefficient times 0.90 = $25.38.

Table 5.6. Cattle parasite treatment determination,
recommended treatments, cost per treatment, and
annual cost per cow unit, Florida, USA
(continued)

Type of parasite	Treatment determination	Cost Per Treatment		
		mature	replacement	calves
Internal				
Liver flukes	Fluke-only areas	2.50	1.80	-
Intestinal				
worms[a]	High nutrition	0.00	1.80	-
	Medium nutrition	2.50	1.80	-
	Low nutrition	2.50	1.80	1.20
Grubs	All Florida	0.60	0.50	-
External				
Horn flies[b]				
Spray[c]	50 or more flies/animal	0.29	0.29	-
Dust bags[d]	50 or more flies/animal	1.24	1.24	-
Ear tags	50 or more flies/animal	1.50	1.50	-
Lice[e]	Presence of lice	0.32	0.32	-

continued

Analyses of net benefits from production practices typically have the annual cost subtracted from the annual benefits to calculate the net benefit per unit. However, because the benefits, although reasonable and defensible, are almost always going to be speculative, this approach should not be taken. Another reason is the wide variability between cattle operations. Thus, benefits are better considered as potential losses in income. In other words, the calculated benefits can be thought of as opportunity losses or potential income that is foregone from not carrying out the practice.

The various dollar benefits for the three categories--mature animals, heifers, and calves--for each of the three areas of potential loss are summarized in Table 5.9 to arrive at an annual potential income loss.

Table 5.6. Cattle parasite treatment determination, recommended treatments, cost per treatment and annual cost per cow unit, Florida, USA (continued)

Type of parasite	Treatment determination	Annual Cost		
		mature	replacement	calves
Internal				
Liver flukes	Fluke-only areas	2.50	1.80	-
Intestinal worms[a]				
	High nutrition	0.00	1.80	-
	Medium nutrition	2.50	3.60	-
	Low nutrition	5.00	3.60	1.20
Grubs	All Florida	0.60	0.50	-
External				
Horn flies[b]				
Spray[c]	50 or more flies/animal	1.74	1.74	-
Dust bags[d]	50 or more flies/animal	1.24	1.24	-
Ear tags	50 or more flies/animal	3.00	3.00	-
Lice[e]	Presence of lice	0.64	0.64	-

[a]Fluke treatments also control most intestinal worms. No additional labor charge assumed.

[b]Other methods such as pour-ons or back rubbers are also used.

[c]Sprayer costs $500 and lasts 5 years = $100 year and with 1,000 cows = 0.10 cow/year = 0.02 per treatment. Labor cost is 250 for 1,000 cows per treatment and 4 treatments in addition to 2 regular workings = $1,000 for 1,000 cows = $1.00 per cow per year = $0.17 per treatment. Materials cost is $0.10 per cow per treatment.

[d]Dust bags. One half kg of dust per head per season. The kits cost $26.50 and have 5.7 kg of dust. Additional dust costs $17.00 for 23 kg. The cost of the kit and dust is $1.04 per head. Also, cost of $0.20 per cow per year is for fence and all associated labor.

[e]Has no labor charge as cattle can be sprayed along with other activities.

Table 5.7. Estimated net benefits from recommended parasite
control procedures in percentage and weight
terms, Florida, USA

Type of parasite	Matures		Replacements		Calves	
	High	Low	High	Low	High	Low
- Net Reduction Death Loss, Percent-						
Internal						
Liver flukes	4	2	4	2	0	0
Intestinal worms						
High nutrition	0	0	0	0	0	0
Medium nutrition	2	0	2	0	0	0
Low nutrition	5	2	5	2	2	0
Grubs	0	0	0	0	0	0
External						
Horn flies	0	0	0	0	0	0
Lice	0	0	0	0	0	0
- - - - - Weight Loss, Kilos - - - -						
Internal						
Liver flukes	45	18	45	18	5	0
Intestinal worms						
High nutrition	9	0	9	0	5	0
Medium nutrition	18	7	18	7	9	5
Low nutrition	27	14	27	14	23	11
Grubs	9	5	9	5	0	0
External						
Horn flies	23	9	23	9	14	5
Lice	14	5	14	5	9	5
Reproductive Loss, Calf Crop, Percent						
Internal						
Liver flukes	12	6	12	6	-	-
Intestinal worms						
High nutrition	0	0	0	0	-	-
Medium nutrition	6	3	6	3	-	-
Low nutrition	12	6	12	6	-	-
Grubs	2	0	2	0	-	-
External						
Horn flies	8	4	8	4	-	-
Lice	5	2	5	2	-	-

Source: Simpson, Kunkle, Sand and Cope, 1984

Table 5.8. Estimated financial benefits from parasite control procedures, Florida, USA

Type of parasite	Mature[a]		Replacements[b]		Calves[c]	
	High	Low	High	Low	High	Low
	- - - - Dollars per cow unit - - - -					
			Death Loss[d]			
Internal						
Liver flukes	13.32	6.66	3.31	1.66	0	0
Intestinal worms						
High nutrition	0	0	0	0	0	0
Medium nutrition	6.66	0	1.66	0	0	0
Low nutrition	16.65	6.66	4.14	1.66	3.53	0
Grubs	0	0	0	0	0	0
External						
Horn flies	0	0	0	0	0	0
Lice	0	0	0	0	0	0
			Weight Loss[e]			
Internal						
Liver flukes	3.50	1.40	1.35	0.54	3.42	0
Intestinal worms						
High nutrition	0.70	0	0.27	0	3.42	0
Medium nutrition	1.40	.53	0.54	0.20	6.84	3.42
Low nutrition	2.10	1.05	0.81	0.41	17.10	8.55
Grubs	0.70	0.35	0.27	0.13	0	0
External						
Horn flies	1.75	0.70	0.68	0.27	10.26	3.42
Lice	1.05	0.35	0.41	0.13	6.84	3.42
			Reproductive Loss[f]			
Internal						
Liver flukes	25.38	12.69	2.82	1.41	3.42	-
Intestinal worms						
High nutrition	0	0	0	0	-	-
Medium nutrition	12.35	6.35	1.41	0.71	-	-
Low nutrition	25.38	12.69	2.82	1.41	-	-
Grubs	4.23	0	0.47	0	-	-
External						
Horn flies	16.92	8.46	1.88	.94	-	-
Lice	10.58	4.23	1.18	.47	-	-

This is done for both the high and low levels. Then the sum of the potential income loss is divided by the cost to arrive at a benefit risk/cost ratio. High ratios indicate more potential benefit or, alternatively, the potential risk if the practice is not followed.

The term benefit risk/cost ratio is used because the more common term, benefit/cost ratio, cannot be properly applied here because it is used in project analysis and carries with it the connotation that discounted (long-term) net benefits are divided by discounted costs. In addition, standard use assumes some initial major capital investment is made in benefit/cost analyses. These problems are discussed in a later chapter on project analysis.

The high, or largest, probable ratio for controlling flukes in mature cows and heifers in fluke areas is 16.4 to 1, while the low side is 8.0 to 1. This means that even the most conservative estimate in this analysis places risk at 8 times more than costs. Thus, considering the high incidence of fluke infestation, this practice will usually return much more than the costs. Alternatively, the benefits are high relative to cost.

The results of treating cows and heifers for intestinal worms definitively show that the poorer the nutrition level, the more benefit from treating for worms. Thus, in years when feed is short, analysts would recommend that special care be taken to carry out a good worming program. Likewise, cattle on poor-quality pasture should be systematically wormed to obtain the highest possible benefits.

Economics of Anabolics

Anabolics, also called implants or growth promotants, are primarily used in cattle and to a much lesser extent in sheep; only minimal use of anabolics is made in goats. Anabolics are also being developed for swine. This example uses beef cattle for all examples, but the analytical method can be used for other types of livestock.

Anabolics such as IMC's Ralgro are employed in a variety of situations including suckling calves not intended for breeding purposes (most anabolics are not cleared for use in breeding stock), growing cattle, (after weaning but prior to the final fattening phase), and during that latter finishing period. Finishing cattle, (those being fattened for slaughter) can be further divided into two types, those on pasture and those in a confinement feedlot.

There are a number of potential physical benefits that depend on the type of cattle and system. Table 5.10 contains a summary of typical benefits found around the world. There are, of course, considerable response differences depending on cattle condition, feed quality and quantity, and climatic variations, just to mention the major factors. However, the range provided, such as 0.088-0.135 kg of additional gain in suckling calves, is expected under usual conditions. The averages provided, such as the 0.111 kg of additional gain in suckling calves, typical of results from hundreds of on-farm trials, are used to calculate the economic benefits.

Table 5.9. Benefit risk/cost ratios per cow unit for
            various parasite treatments, example problem

Type of parasite	Annual cost[a]	Annual potential income loss[b]		Benefit risk/cost ratio	
		High	Low	High	Low
	- - - - - Dollars per cow unit - - - -				
Internal					
Liver flukes	3.04	49.68	24.36	16.4:1	8.0:1
Intestinal worms					
High nutrition	0.41	4.39	0.00	10.7:1	--
Medium nutrition	4.36	30.86	11.21	7.1:1	2.6:1
Low nutrition	6.98	72.53	32.43	10.4:1	4.6:1
Flukes and worms[c]					
High nutrition	6.08	54.07	24.36	8.9:1	4.0:1
Medium nutrition	6.08	80.54	35.57	13.2:1	5.9:1
Low nutrition	6.08	112.21	56.79	20.1:1	9.3:1
Grubs	0.75	5.67	0.48	7.6:1	0.6:1
External					
Horn flies					
Spray	2.23	31.49	13.79	14.1:1	6.2:1
Dust bags	1.59	31.49	13.79	19.8:1	8.7:1
Ear tags	3.84	31.49	13.79	8.2:1	3.6:1
Lice	0.82	20.06	8.60	24.5:1	10.5:1

Source: Simpson, Kunkle, Sand and Cope, 1984.

[a]All costs are in $/cow unit. A cow unit is 1.05 mature
animals (to account for bulls), 0.23 replacement heifers (1
and 2 year heifers) and 0.75 calves. For example, taking
data from Table 5.6, liver flukes only are $2.50 times 1.05
= $2.63 plus $1.80 times 0.23 = $0.41 totaling $3.04.

[b]Benefits are from Table 5.8 by summing death, weight and
reproductive loss for matures, replacements and calves.
For example, $13.32 + $3.31 + $3.50 + $1.35 + 25.38 + $2.82
= $49.68 as the high annual potential income loss from
liver flukes.

[c]Control of worms as well as flukes is obtained by using
some commercial flukacides.

Table 5.10. Physical benefits from one 90-day cattle implant

Benefit	Suckling Calves	Growing Cattle	Pasture Finishing Cattle
Typical weight			
Range	Birth-200	200-360	360-550
Average	100	280	455
Additional average daily gain (kg)			
Range	.088-.135	.111-.178	.133-.222
Average	.111	.155	.178
Additional average daily gain (pct)			
Range	10-14	12-15	15-20
Average	12	14	16
Additional total gain (kg)			
Range	8-12	10-16	15-20
Average	10	12	15
Additional total gain (pct)			
Range	4.0-6.0	2.8-4.4	2.7-3.6
Average	5.0	3.3	3.1

Source: Simpson, 1987.

One principal benefit attributable to anabolics in all classes of cattle is additional gain, which may be measured in at least three ways: additional average daily gain as calculated in kilos, additional average daily gain as calculated in percentage terms, and total additional daily gain.

Partial analysis is the proper approach to measure benefits from anabolics. The first examples include six methods in suckling calves; the choice of method depends on its relevance to an individual producer, (Table 5.11). However, in all cases the analysis centers on additional gain obtained from using an implant.

The first method is a calculation of return per implant using the typical values of 10 kg of additional gain per implanted suckling calf. The analysis indicates that although a typical cost per implant is about $1.40, return would be about $12.50 if calves were priced at $1.25 per kg, thus providing a net return to management of $11.10. That translates, as

Table 5.11. Suckling calf economic analyses

Item	Parameter[a]
Method 1 Return per Implant	
Cost	
1 implant ($)	1.25
1/500 implant gun @ $20 each ($)	0.04
1 minute @ $4.00/hr ($)	0.07
Misc. expenses ($)	0.04
Total	1.40
Return	
10 kg @ $1.25 per kg ($)	12.50
Net return to management ($)	13.00
Method 2 Return per $1.00 Invested	
Return per implant period	
(10 kg @ $1.25 per kg) ($)	12.50
Cost per implanted animal	
(See method 1) ($)	1.40
Return per $1.00 invested	
(Return divided by cost) ($)	8.93
Method 3 Annual Return per $1.00 Invested	
Implant period (days)	90
365 days minus implant period (days)	275
Remaining period as fraction of year	.75
Opportunity value on money (pct)	8
Equivalent opportunity value,	
remaining period of year	6
Return per $1.00 invested, end of 90 days ($)	8.93
Return per $1.00 invested, end of year	
($8.93 times 1.06) ($)	9.47

continued

Table 5.11.   Suckling calf economic analyses (continued)

Item	Parameter[a]

Method 4 Additional Gain Required to Pay Implant Cost

Total cost ($)	1.40
Value per Kilo ($)	1.20

Kilos required per 90
  day period (cost divided
  by value per kg) (kg)   1.12

Method 5 Return per Hectare

Cows per hectare (no)   0.50

Calf crop (pct)   70

Net Return per implant per calf
  (See Method 1) ($)   11.10

Return per hectare per implant
  (Return per implant per calf
    times calves per hectare) ($)   3.89

Implants per calf per year (no)   2

Annual net additional return per hectare
  from implant ($)   7.78

Method 6 Annual Return per Hectare per $1.00 Invested

Calves per hectare (no)   .35

Gross return per calf per implant ($)   12.50

Number of implants (no)   2

Total gross return per hectare ($)   8.75

Additional cost (cost per implant times number
  of implants times calves per hectare) ($)
  ($1.40 x 2) (0.35)   .98

Return per $1.00 invested ($)
  (Gross return divided by additional cost)   8.93

Source: Simpson, 1987.
[a]U.S.$ in all cases

shown in method two, to $8.93 for every dollar invested--and only during a 90-day period. If, as analyzed in method three, the evaluation is placed on a 1-year basis, the return is $9.47. An appropriate question for the cow/calf producer then is: Do I have alternative investments on my farm or ranch that will yield me more than $9.47 annually for each $1.00 invested. If so, then those investments should have priority over implants.

Another method to evaluate the financial soundness of implants is to calculate the additional gain required to pay back the cost. The results in method four indicate that if calves were priced at $1.25 per kg, and the cost were $1.40 per implant, then only 1.12 kilos of additional gain would be required to break even.

Return per hectare is the fifth method provided in Table 5.11. If carrying capacity were 1 cow on 2 hectares (0.50 cows per ha), a 70 percent calf crop obtained, and a net return per implant of $11.10 (as calculated earlier), then the net return per hectare would be $3.89 per implant. Calves can be implanted the first month after birth, so suckling calves could also be reimplanted prior to weaning. The net result would be a return of $7.78 per hectare.

Another approach is to calculate the return per hectare for each $1.00 invested. Results, shown as method 6, indicate that $8.93 would be gained, which is the same as in method two because the analysis is on a constant-unit basis. Similarly, the return per dollar invested would also be the same if placed on a brood cow basis.

The type of analysis for growing and pasture finishing cattle is the same as for suckling calves. Computations are not made, but results are provided in Table 5.12.

Table 5.12.   Summary of net return from anabolics in suckling calves, growing cattle and pasture cattle

Class of cattle	One implant	$1.00 invested	Return per annual per $1.00 invested	Hectare	Additional gain to breakeven
	- - - - - - - - - - U.S.$ - - - - - - - - - -				- kg -
Suckling calves	13.00	8.93	9.47	7.78	1.12
Growing cattle	13.00	10.29	10.91	52.00	1.17
Pasture finishing	13.60	10.71	11.35	54.40	1.40

Source: Simpson, 1987.

CHAPTER 6

ECONOMIC ANALYSES IN FATTENING CATTLE
TO SLAUGHTER WEIGHT

Several alternatives for fattening cattle to slaughter weight were presented in Chapter 4 to demonstrate ways in which budgeting can be used in this type of problem. The subject of feeding cattle to slaughter weight is now considered in more detail; the main emphasis here is on problems faced by developing countries in determining the feasibility of various feeding systems.

Types and Economics of Feeding Facilities

There are an almost infinite number of facilities in which cattle can be finished to slaughter weights, ranging from those with virtually no capital investment to ones that are highly sophisticated with total environmental control. Pasture finishing requires relatively little investment in facilities, but may entail a substantial capital outlay in forage development and fencing for confinement feedlots. There is no one type of facility that is best; the optimal type may be radically different even between neighboring feeders. One individual or group might have a strong financial position, feed cattle all year and have an aversion to mud. That producer might then find that a well-built barn with a floor made of concrete slats to permit manure to drop through to a sump, where the manure is removed by a mechanical scraper and then spread on fields by a sprinkler system, to be the most economic and personally most appealing. On the other hand, another feeder, who might feed cattle only on an irregular basis, may not be in a financial position that can accommodate the debt from construction of a capital intensive facility and may not even have the management ability to handle more than the most rudimentary operation. Consequently, the optimal strategy for that person might be homemade open troughs in a barbwire enclosure. Comparison of feeding systems is examined in the next section, and results of feeding trials using capital-intensive methods are compared with growing and finishing cattle under minimum capital investment conditions.

Methods for Carrying Out Economic Analyses
of Beef Cattle Feeding Trials

Results from beef cattle growing and finishing feeding trials are often reported in the literature, but no generalized methods have been presented to convert the physical data into economic analyses. The purpose of this section is to present an appropriate method that will permit analysts to quickly determine the economic feasibility for commercial feeders who use a ration or feeding approach reported by experiment stations or other researchers. Silage is chosen as an example because it has received considerable attention worldwide as a partial or complete substitute for grain.

Overview of Procedure

A number of approaches are possible to determine a feeding or forage trial's economic viability. However, it appears that the most useful one would answer the question: Is this feeding procedure or ration more cost effective than alternatives? In order to quantify the results there must be a standard or norm--in effect, a procedure for making both intra-and interfirm comparisons. Consequently, the data should be presented in a format that applies to a specific area but that can be modified to fit other situations.

The problem of fluctuation in output and input prices can be overcome by comparing the trial results to a localized industry norm and then presenting the results in a fashion that permits potential adopters of the new practice to evaluate the results using their own economic and physical relationship data. Based on this method of comparing trial results, the procedure is

1. Develop a table of cattle performance in the feedlot trials.
2. Calculate the ration in kg per head per day, dry matter basis.
3. Develop a table of present ingredient costs on an as-feed and dry matter basis.
4. Calculate the cost per head per day for the ration.
5. Calculate cost per kg of gain for the ration only.
6. Calculate cost per kg of gain in a complete budget.
7. Compare cost per kg of gain from the trial with cost from current practices.
8. Analyze the results and determine any constraints to adoption of the new practice, providing it is shown to be more profitable.

Rations and Performance

The following feeding trial assumes that 50 steers in a trial average 216.4 kg when started on the growing phase 17 days after they arrive at the experiment station. The finishing phase is begun 103 days after that.

The cattle are divided into five lots of ten head each. The growing ration is varied from a high of 75 percent silage on a dry matter basis to a low of 2 percent (Table 6.1). The average daily gain varies from 0.87 kg per head for the high silage ration to 1.37 kg for the low silage fed cattle. Total dry matter (DM) fed declines from 3.36 kg per kg of gain with the high silage ration to 2.23 kg with the low silage ration.

Cattle in the finishing phase are held on feed for a longer time (a maximum of 119 days) on the high silage ration than on the low silage one (70 days). Average daily gain varies from 1.04 kg per head with the high silage ration to 1.29 kg per day with the low silage ration. Total dry matter fed declines from 3.85 kg per kg of gain to 2.61 per kg of gain with the low silage operation.

Economic Analysis

The ration in kg per head per day for the growing and finishing phases, on a dry matter basis, is presented in Table 6.2. The ingredient costs, given in Table 6.3, are first provided on an as-fed basis and then divided by the percentage of dry matter to arrive at a cost per unit of dry matter. The units, such as tons, are then converted to a per kg basis in the last column. Feedlot operators would have to use dry matter content of their own ingredients to calculate costs.

The cost per head per day for the feed only, and on a dry matter basis, calculated in Table 6.4., increases from $0.73 for lot 1 in the growing phase (75 percent silage on a dry matter basis) to $1.12 in the low (2 percent) silage ration. The cost per head increases continually as maize constitutes a higher percent of the ration. The finishing phase cost per head per day varies considerable, from $1.11 in the high silage ration (57 percent of DM) to $1.27 when there is no silage in the ration.

Totals for the combined phases are given in Table 6.5. The average cost per head per day is lowest for the all-silage ration ($0.93) and highest ($1.18) for the high maize ration. However, the cost per kilo of gain is just the opposite, for it declines as the percentage of maize increases. On an overall basis for feed only, the cost is $0.97 per kg of gain for lot 1, the high silage ration, compared with $0.88 for number 5, which is high maize. The lowest cost per kg of gain in the growing phase is for the medium silage ration (number 3), indicating that overall, probably the most economic situation would be the ration fed to lot 3 in the growing phase and the ration fed to lot 5 in the finishing phase.

Complete Feedlot Budget

The preceding partial budgeting analysis should be sufficient to make a determination of the optimal ration from the feedlot trial. But at times, even though a new feeding practice may appear to be lucrative, there may be other mitigating circumstances whose impact should be evaluated using a complete feedlot budget. An example of this aspect is now provided using the problem developed in Chapter 4.

The growing and grain finishing phases from the original budgets in Chapter 4 are integrated in Table 6.6 to calculate costs and returns as if the operation were a continual one. It can be recalled that this feeding

Table 6.1. Feedlot performance, feeding/economic method
            example

Item	Growing phase by lot number				
	1	2	3	4	5
Days	103	103	103	103	103
Maize					
Dry matter (%)	10	31	43	56	85
Gross wt. (%)	6	20	30	42	85
Silage					
Dry matter (%)	75	57	45	33	2
Gross wt. (%)	87	74	64	51	3
Cattle					
Initial wt. (kg)	215.91	216.36	217.73	215.91	215.91
Final wt. (kg)	305.45	335.91	354.55	350.91	357.27
Avg. daily gain (kg)	0.87	1.16	1.33	1.31	1.37
Feed/head/day					
Dry matter (kg)	6.41	6.95	7.50	7.50	6.77
Gross wt. (kg)	15.32	15.05	14.91	13.64	9.14
Kg feed dry matter/kg gain					
Total (kg)	7.39	6.03	5.66	5.73	4.90
Corn (kg)	0.76	1.84	2.45	3.18	4.17
Silage (kg)	5.53	3.44	2.55	1.91	0.11

continued

period begins immediately after weaning and continues for 285 days; cattle are sold at a slaughter weight of 461 kg. Some items in the combined budget are not additive between the growing and finishing phases. These are highlighted in the footnotes.

Changing the operation to conform with the experimental practice means the weaned steer calves are placed directly in a confinement facility for the growing phase rather that being grazed on bahia and ryegrass pasture. The steers are assumed to weight 205 kg (same as the current practice) and must be fed 113 days to reach the final weight of 355 kg, at which time the steers enter the finishing phase. The ration and all assumptions relating to it are the same as experimental lot number 3 for the growing phase and lot number 5 for the finishing phase. Care must be taken in feed cost computation because ration requirements and costs are on a dry matter basis whereas the current practice was

Table 6.1. Feedlot performance, feeding/economic method
example (continued)

Item	Finishing phase by lot number				
	1	2	3	4	5
Days	119	102	80	70	70
Maize					
Dry matter (%)	35	54	71	81	91
Gross wt. (%)	22	39	59	74	92
Silage					
Dry matter (%)	57	38	21	10	0
Gross wt. (%)	73	56	35	19	0
Cattle					
Initial wt. (kg)	305.46	335.46	354.55	350.46	357.28
Final wt. (kg)	429.55	448.19	431.82	437.73	447.28
Avg. daily gain (kg)	1.04	1.11	0.97	1.25	1.29
Feed/head/day					
Dry matter (kg)	8.64	8.50	8.05	8.05	7.50
Gross wt. (kg)	18.69	16.09	13.09	12.00	10.00
Kg feed dry matter/kg gain					
Total (kg)	8.46	7.70	8.76	6.42	5.74
Corn (kg)	2.97	4.15	6.20	5.21	5.20
Silage (kg)	4.83	2.92	1.81	0.66	0.00

calculated on an as-fed basis. The cattle are in the finishing phase for 82 days with a final weight of 461 kg, the same as in the current practice.

Total time on feed is 195 days or 6.5 months. Because the cattle are sold before the hot rainy season, no shades are necessary in the feedlot that must be constructed to hold the cattle (in contrast to the relatively primitive facilities if cattle are only fed a short period). This is a major savings because shades are one of the most costly items in a feedlot (Simpson, Baldwin and Baker, 1981). However, a facility to hold high moisture maize (which can also be used for silage) must be constructed. The net result is a total investment of $58,834 in comparison with $23,474 under current practices.

Production costs are lower with the proposed practice because the conversion with high moisture maize is about 10-12 percent better than

Table 6.2. Example ration in kilos per head per day for the growing and finishing phases, dry matter basis

Ingredient	Lot number				
	1	2	3	4	5

- - - - - - - Kg/hd/day - - - - - - -

Growing phase

Ingredient	1	2	3	4	5
Maize silage	4.82	4.00	3.41	2.50	0.09
High moisture maize (shelled)	0.46	1.91	3.05	3.96	5.55
Protein supplement (61%)	0.69	0.64	0.64	0.64	0.64
Cane molasses	0.23	0.23	0.23	0.23	0.23
Hay	0.23	0.19	0.19	0.19	0.19
Total	6.41	6.96	7.50	7.50	6.78

Finishing phase

Ingredient	1	2	3	4	5
Maize silage	4.91	3.23	1.64	0.82	0.00
High moisture maize (shelled)	2.78	4.32	5.46	6.23	6.50
Protein supplement (56%)	0.69	0.69	0.69	0.69	0.69
Cane molasses	0.28	0.28	0.28	0.28	0.28
Hay	--	--	--	--	--
Total	8.64	8.50	8.05	8.05	7.50

with whole shell maize and less time is required on feed in the growing phase. This advantage is offset somewhat by other higher costs, mainly due to more interest, to opportunity costs charged to the additional investment, or, especially, to care required in management of the storage facility. Furthermore, feeding high moisture maize on a small scale is quite a complex undertaking because the maize must be stored for relatively long periods in an airtight container. The one low cost possibility examined here, air bags made from heavy-duty plastic, present considerable practical problems, such as protection from wild animals. This is another example of the need to evaluate management considerations.

The result of all calculations is net income above production costs amounts to $20,147, which is about 2.5 times the current practice. Furthermore, even other costs such as management, ownership costs and capital costs are covered, whereas they are not met in the current practice. The breakeven costs are much lower, as are the costs of gain.

Table 6.3. Ingredient costs, as-fed and dry matter basis, example problem

Ingredient	Cost per ton as-fed basis	Percent dry matter	Cost per kg dry matter basis
Maize silage	33.48	36	0.0918
High moisture maize (shelled)	119.88	74	0.1620
Protein supplement (56%)	220.00	90	0.2445
Cane molasses	100.00	76	0.1448
Hay	66.00	90	0.0733

Table 6.4. Cost per head per day for growing and finishing phases, dry matter basis, example problem[a]

Ingredient	Lot number				
	1	2	3	4	5
	- - - - - - - $U.S. - - - - - - -				
	Growing phase				
Maize silage	.442	.367	.313	.230	.009
High moisture maize (shelled)	.074	.310	.494	.641	.898
Protein supplement (61%)	.167	.156	.156	.156	.156
Cane molasses	.033	.033	.033	.033	.033
Hay	.017	.014	.014	.014	.020
Total	.733	.880	1.010	1.074	1.116
	Finishing phase				
Maize silage	.451	.296	.151	.075	.000
High moisture maize (shelled)	.449	.700	.884	1.009	1.053
Protein supplement (61%)	.167	.167	.167	.178	.178
Cane molasses	.040	.040	.040	.040	.040
Hay	.000	.000	.000	.000	.000
Total	1.107	1.203	1.242	1.302	1.271

[a]Computed from data in Tables 6.2 and 6.3.

Table 6.5. Total cost of ration and cost per kg of gain, ration only, example silage versus grain-feeding problem

Ingredient	Lot number				
	1	2	3	4	5
Growing phase					
Average daily gain (kg)[a]	0.87	1.16	1.33	1.31	1.37
Total days[a]	103.00	103.00	103.00	103.00	103.00
Total gain (kg)[b]	89.55	119.55	136.82	135.00	141.36
Cost per head per day (U.S.$)[c]	0.733	.880	1.010	1.074	1.116
Total cost per head (U.S.$)[d]	75.50	90.64	104.03	110.62	114.95
Cost per kg of gain (U.S.$)[e]	0.84	0.77	0.75	0.81	0.81
Finishing phase					
Average daily gain (kg)[a]	1.04	1.11	0.97	1.25	1.29
Total days[a]	119.00	102.00	80.00	70.00	70.00
Total gain (kg)[b]	124.09	112.73	77.27	87.27	90.00
Cost per head per day (U.S.$)[f]	1.107	1.203	1.242	1.302	1.271
Total cost per head (U.S.$)[d]	131.73	122.70	99.36	91.14	88.97
Cost per kg of gain (U.S.$)[e]	1.06	1.10	1.30	1.06	0.99
Combined phases					
Average daily gain (kg)[g]	0.96	1.13	1.17	1.29	1.34
Total days	222.00	205.00	183.00	173.00	173.00
Total gain (kg)	213.64	232.27	214.55	222.27	231.82
Cost per head per day (U.S.$)[h]	0.93	1.04	1.11	1.17	1.18
Total cost per head (U.S.$)	207.23	213.34	203.39	201.76	203.92
Cost per kg of gain (U.S.$)[e]	0.97	0.92	0.95	0.90	0.88

[a]From Table 6.1.
[b]Calculated from Table 6.1.
[c]From Table 6.4.
[d]Total days times cost per day
[e]Total cost per head divided by total gain.
[f]From Table 6.4

[g]Summation of total gain from both phases divided by summation of days, both phases.
[h]Total cost per head, both phases, divided by total days.

Table 6.6. Comparison of current practices and adoption of
method used in experimental trials, example
problem

Item	Current Practice[a] Growing	Grain finishing	Combination	Combination experimental practice[b]
	- - - - - - - $U.S. - - - - - - -			
Investment				
Land	0	1,480	1,480	1,960
Buildings & capital investment	9,500	9,000	18,500	23,630
Equipment	2,270	1,224	3,494	33,244
Cattle	0	0	0	0
Total	11,770	11,704	23,474	58,834
Basic production costs				
Cattle	30,750	46,654	30,750	27,060
Pasture or silage	6,101	0	6,101	3,537
Hay	405	1,360	1,765	157
Grain mix or high moisture corn	3,375	21,384	24,759	12,836
Salt or protein supplement	25	6	31	3,151
Minerals or molasses	77	0	77	1,196
Medicine	213	388	601	601
Veterinary services	0	50	50	50
Maintenance, fuel, repairs	500	106	606	1,174
Labor, hired	0	0	0	0
Other & administrative	200	175	375	375
Total, without cattle	14,586	23,469	38,055	23,077
Total, with cattle	41,646	70,123	68,805	50,137
Other Costs				
Capital costs				
Land	--	178	178	178
Buildings & capital investment	1,140	1,080	2,220	1,990
Equipment	272	147	419	200
Livestock	0	0	0	0
Operating capital	2,624	7,573	10,197	10,197
Subtotal	4,036	8,978	13,014	12,565

continued

Table 6.6. Comparison of current practices and adoption of method used in experimental trials, example problem (continued)

| Item | Current Practice[a] | | | Combination experimental practice[b] |
	Grow-ing	Grain finishing	Combina-tion	
	- - - - - - - $U.S. - - - - - - -			
Ownership costs				
Depreciation	702	572	1,274	2,844
Taxes	235	234	469	1,777
Insurance	<u>117</u>	<u>117</u>	<u>234</u>	<u>588</u>
Subtotal	1,054	923	1,977	5,209
Land rent	0	0	0	0
Labor, owners	956	510	1,466	829
Management	1,500	1,500	3,000	3,000
Total, other costs	7,546	11,911	19,457	21,603
Total, basic production & other costs, no cattle	22,132	35,380	57,512	44,680
Total, basic production & other costs, with cattle	49,192	82,034	84,572	71,740
Income	46,654	69,574	69,574	70,284
Net income				
Above net production costs	5,008	549	769	20,147
Above all costs	-2,538	-12,460	-14,998	-1,456
Breakeven				
Production costs	1.29	1.55	1.52	1.10
All costs	1.53	1.82	1.87	1.57
Cost of gain				
Production costs	1.25	1.80	1.54	0.92
All costs	1.90	2.72	2.33	1.78

[a]Growing and grain finishing from Chapter 4. Combination calculated.
[b]Calculations not shown.

The analysis indicates that producers would be economically much better off shifting to the system reported by the experiment station unless there are other mitigating circumstances, such as only feeding cattle every few years. The major point of the exercise is to demonstrate the way in which experiment station results can be interpreted from an economic viewpoint and, then, how the results can be used to analyze the impact on an operator's own program. Both enterprise and partial budgeting methods were utilized. Now the method of partial budgeting is extended to a further analysis of anabolics.

Anabolics and Ionophores in Feedlot Cattle

The economic analysis on anabolics using the partial budgeting method given in Chapter 5 for calves, growing cattle and pasture finishing cattle is now extended to confinement feedlot situations. Research indicates that improved feed efficiency or conversion of feedstuffs to body development, measured in percentage terms, is another benefit in feedlot finishing cattle. There are two possibilities, improved efficiency when anabolics are used alone and even greater efficiency when anabolics are combined with an ionophore. There is improved feed efficiency from using ionophores in growing and finishing cattle on pasture, but due to measurement difficulties (for example, due to forage type and seasonal impact), published data are not available to document this benefit. Consequently, even though it is probable that substantial benefits are derived from using ionophores in pasture cattle, typical benefits were not provided in Table 5.10 for this production practice.

There are several benefits from anabolics apart from weight gain. For example, it has been shown that implanted cattle have improved stress tolerance. These benefits are reflected in lower death loss when cattle are shipped and also in improved general health. Data are not available on direct benefits to producers from reduced death loss, so this factor is not quantified in the economic analysis. Improved health is reflected in additional average daily gain.

Fattened cattle (also called fed or slaughter cattle) can be sold at an earlier age with a given target weight due to faster daily gain. This is of special interest to confinement feedlot operators who want to move cattle in and out of lots as quickly as possible. Alternatively, implanted cattle fattened for the same period of time as nonimplanted cattle will have heavier carcasses. This latter benefit is important to producers fattening cattle on pasture, especially where there are distinct wet and dry seasons with consequent fixed amounts of forage available.

The approach to calculate benefits from using anabolics in feedlot cattle is quite different from the one used in the previous three production systems. Cost saving rather than additional gain is now the focal point. This is because feedlot cattle are usually fed to a specified grade that analytically can be considered a given weight, depending on the breed.

The benefits fall in two classes, feed efficiency improvement and reduced interest on money invested. The first part of the feed efficiency portion is to estimate total feed use and associated costs under a base situation in which neither an implant nor an ionophore is used. Then the reduced days on feed as a result of greater average daily gain are calculated. After this, the new feed consumption figures resulting from greater feed efficiency are calculated.

Implants alone provide an additional 7.5 percent average improvement in feed efficiency (as shown in Table 6.7). This increases to 10 percent improvement when cattle are reimplanted. Efficiency is improved by 14 percent when there is one implant used in combination with an ionophore. That efficiency level increases to 16 percent when an ionophore is combined with a second implant. Thus, four separate analyses must be made.

The net result of calculating reduced feed cost, reduced feed handling and management cost, and reduced interest or opportunity cost on money invested in cattle and feed is a net benefit of $24.72 with one implant alone and $41.63 with two implants (Table 6.8). Benefits are only slightly higher when an ionophore is added, $29.30 with one implant and $47.94 when two implants are used.

The annualized return per $1.00 invested is $19.06 for one implant alone, but $10.82 when used in combination with an ionophore. The return per dollar invested is lower with the ionophore because this is a marginal concept. The economic rule, as discussed in Chapter 3, is to continue adding inputs until the return falls to the input cost, that is, where the MIC is equal to the MVP. At that point, total profit will be highest. Indeed, the total return from the combination is higher than with a growth promotant alone. The annualized return per dollar invested from two implants alone is $15.96 while it is $10.09 when used in combination with an ionophore.

There are other benefits from using implants in feedlot cattle that are not included in the analysis. For example, reduced periods on feed permit increased marketing flexibility. Cattle can be held a somewhat longer period if needed. The time factor is also a major variable to many other producers. For example, cattle grazed to slaughter weight in the tropics will often be held to four, and certainly to three, years of age. A major problem is alternating wet/dry seasons. If anabolics are used, cattle that would normally reach slaughter weight at 4 or perhaps 3 years of age can reach that weight more quickly, often by one year less. When that occurs, producers can stock 25-30 percent more mature animals. Those adults could either be cows or finishing cattle. Thus, there are considerable benefits in this instance not given in the economic analysis.

Table 6.7. Physical benefits from feedlot cattle implants
           and ionophores

Benefit	Parameters
Typical weight (kg)	
Range	360-525
Average	440
Additional average daily gain (kg)	
Range	.155-.244
Average	.200
Additional average daily gain (pct)	
Range	15-20
Average	16
Additional total gain (kg)	
Range	14-22
Average	16
Improved feed efficiency (conversion) (pct)	
Implant alone (1)	
Range	7-10
Average	7.5
Reimplant (2nd one)	
Range	9-11
Average	10
With ionophore, initial implant	
Range	13-15
Average	14
With ionophore, reimplant (2nd one)	
Range	15-17
Average	16

Source: Simpson, 1987.

Feasibility Studies

Feeding cattle to slaughter weight in confinement feedlots is a specialized business that is considerably different than raising calves or pasture finishing cattle, mainly because of the need for substantially more intensive management. In effect, a cattle feedlot bears more similarity to

Table 6.8. Feedlot cattle economic analyses

Item	Implant only		Implant & ionophore	
	First implant	First implant & reimplant	First implant	First implant & reimplant
Method 1 Return per Implant				
A. Reduced feed cost due to greater efficiency				
Days on feed (days)	90	140	90	140
Feed conversion w/o Ralgro or ionophore (ratio)	8.6	8.6	8.6	8.6
Average daily gain (kg)	1.35	1.35	1.35	1.35
Feed consumption w/o implant & ionophore				
Daily (kg)	11.61	11.61	11.61	11.61
Total, 90 days (kg)	1045	--	1045	--
Total 140 days (kg)	--	1625	--	1625
Feed cost no implant or ionophore				
Per kg ($)	0.09	0.09	0.09	0.09
Total ($)	94.05	146.25	94.05	146.05
Additional average daily gain (pct)	16	16	16	16
Reduced days on feed (additional ADG pct. times base days) (days)	14	22	14	22
Revised days on feed	76	118	76	118
Feed efficiency improvement w/o implant & ionophores (pct)	7.5	10.0	14.0	16.0

continued

Table 6.8. Feedlot cattle economic analyses (continued)

Item	Implant only		Implant & ionophore	
	First implant	First implant & reimplant	First implant	First implant & reimplant
Feed consumption with implant & ionophores				
Daily (kg)	10.74	10.45	9.98	9.75
Total (kg)	816	1233	758	1151
Feed cost with implant & ionophore				
Per kg ($)	0.09	0.09	0.09	0.09
Total ($)	73.44	110.97	68.22	103.59
Reduced feed cost due to efficiency ($)	20.61	35.28	25.83	42.66

Method 1 Return per Implant

B. Reduced feed handling and management cost

Item	Implant only		Implant & ionophore	
Charge per ton of feed ($)	15.00	15.00	15.00	15.00
Feed reduction (kg)	229	392	287	474
Reduced charge	3.43	5.88	4.31	7.11

C. Reduced interest or opportunity cost on money invested

Interest

Item	Implant only		Implant & ionophore	
Rate (annual) (pct)	10	10	10	10
Rate (daily) (pct)	.027	.027	.027	.027

continued

Table 6.8. Feedlot cattle economic analyses (continued)

Item	Implant only		Implant & ionophore	
	First implant	First implant & reimplant	First implant	First implant & reimplant
Reduced days on feed	14	22	14	22
Interest factor	.00378	.00594	.00378	.00594
Value of				
Feeder cattle (100%) ($)	475	475	475	475
Feed (50%) ($)	75	75	75	75
Total ($)	550	550	550	550
Reduced cost due to interest ($)	2.08	3.27	2.08	3.27
Total reduced cost ($)	26.12	44.43	32.22	53.04
Additional Costs				
Implant ($)	1.25	2.50	1.25	2.50
Implant, 1/500 of cost ($20 per implant) ($)	0.04	0.08	0.04	0.08
1 minute per implant, $4.00 per hour ($)	0.07	0.14	0.07	0.14
Ionophore ($2.00 per ton of feed) ($)	- -	-	1.52	2.30
Misc. expenses ($)	0.04	0.08	0.04	0.08
Total ($)	1.40	2.80	2.92	5.10
Net benefit				
Reduced costs	26.12	44.43	32.22	53.04
Added costs	1.40	2.80	2.92	5.10
Net benefit	24.72	41.63	29.30	47.94
Method 2 Return per $1.00 Invested				
Reduced cost ($)	24.72	41.63	29.30	47.94
Additional cost ($)	1.40	2.80	2.92	5.10

continued

Table 6.8. Feedlot cattle economic analyses (continued)

Item	Implant only		Implant & ionophore	
	First implant	First implant & reimplant	First implant	First implant & reimplant
Return per $1.00 invested (reduced divided by additional)	17.66	14.87	10.03	9.40
Method 3 Annual Return per $1.00 Invested				
Implant period (days)	76	118	76	118
365 days minus implant period (days)	289	268	289	268
Remaining period as fraction of year	.79	.73	.79	.73
Opportunity value on money invested (pct)	10	10	10	10
Equivalent opportunity value (EOV) remaining period of year (pct)	7.9	7.3	7.9	7.3
Return per $1.00 invested, cattle feeding period (from method 2) ($)	17.66	14.87	10.03	9.40
End of year (return times (1 + EOV) ($)	19.06	15.96	10.82	10.09
Method 4 Additional Gain Required to Pay Implant Cost				
Additional cost ($)	1.40	2.80	2.92	5.10
Value live cattle ($/kg)	1.40	1.40	1.40	1.40
Additional gain required (cost divided by value) (kg)	1.0	2.0	2.1	3.6

Source: Simpson, 1987.

a processing industry or factory than the primary production unit does. Thus, even though a good knowledge of cattle is crucial, success is heavily weighted to possessing a factory-type business orientation. Unfortunately, this distinction is often lost on potential entrants to this phase of the livestock business, and consequently the failure rate is fairly high. A major objective of this section is to point out the complexity of cattle feeding and the means to evaluate it as a business.

Government officials, farmers, or other interest groups often decide that some means of finishing cattle should be organized that can serve to improve the current system in order to increase national productivity, reduce costs to consumers or attempt to take some of the "middleman's" profit. Once the specialized nature of the business is recognized and a is decision made to proceed, what might be called a "prefeasibility study," which centers on a few basic questions, should be carried out. The major questions are   1) is there a sufficient supply of cattle?   2) is there a source of relatively cheap feed and   3) is the cost/beef price ratio sufficiently favorable that a complete study is warranted?

Assuming that cattle and feed are deemed available, a quick means to evaluate the cost/beef price ratio problem is to calculate the price of fed cattle needed for the operation to break even. A rapid way to make the computations, shown in Table 6.9, requires data that can be obtained from available publications even in different countries but with similar conditions, through the extension service, at an experiment station or from others engaged in the business. The objective at this point should not be to carry out a detailed analysis, but rather to make some rapid estimates to determine if a complete feasibility study is warranted.

The easiest calculation is determination of fed cattle price required to cover costs of feed and the feeder animal. A typical ration is first set forth, the cost per head per day estimated and the number of days on feed computed. Then total costs are divided by final weight to arrive at the breakeven price, which, using as an example ration number 5 from the finishing phase in Table 6.4, is $1.25 per kg. If these major costs are covered--that is, the probable price of fed cattle is greater than the breakeven price--other production costs such as fuel and repairs, management, labor, and so forth can be added, and a new breakeven can be computed, $1.34 in this case. After that, other items such as capital and ownership costs can be estimated and the final total cost per head divided by final weight to arrive at the production cost per kilo. In this example the finished animals would have to be sold at $1.43 per kg to cover all costs.

Some organizations may not feel that all costs need to be covered, but these organizations probably will have to at least cover basic production costs. An example of this latter case is a slaughter plant that is considering integrating backward to assure a steady supply of slaughter cattle so the plant can continue to operate at full capacity. This is a method of spreading costs. Another example is a government intent on fostering cattle feeding by a subsidy (the infant industry argument).

Table 6.9 Abbreviated method to determine if cattle feeding is feasible

Step		Example		
	Item	Quantity per/hd/day (DM)	Cost/kg (DM)	Total Cost hd/day
Breakeven on Feed Costs		- kg -	- - - $U.S. - - - -	
1) Calculate a typical ration and cost per head per day	Shell corn	6.50	0.1620	1.053
	Protein supp.	0.69	0.2445	0.178
	Molasses	0.28	0.1448	0.040
	Total			1.271

2) Calculate net gain

    Avg. weight in 357.28 kg
    Avg. weight out <u>447.28</u> kg
    Net gain   90.00 kg

3) Estimate the average daily gain   1.29 kg
4) Calculate the number of days on feed   90 kg 1.29 kg = 70 days
5) Calculate total cost of feed   70 days x $1.271 = $88.97
6) Calculate total cost of feeder $1.32/kg x 357 kg = $471.24
7) Calculate total costs $471.24 + $88.97 = $560.21
8) Calculate breakeven price of fed cattle to cover feed and feeder costs   Total cost ($560.21) weight out (447.28 kg) = $1.25/kg

Breakeven on Production Costs

9) Estimate other production costs per head $0.44/day x 90 days = $39.60
10) Calculate total costs $560.21 + $39.60 = $599.81
11) Calculate breakeven price of fed cattle $599.81 477.28 kg = $1.34

Breakeven on all costs

12) Estimate other feeding costs per head $0.42/day x 90 days = $37.80
13) Calculate total costs $599.81 + 37.80 = $637.61
14) Calculate breakeven price of fed cattle   637.61 447.28 kg = $1.43

A complete study can be carried out if the abbreviated, (prefeasibility) study provides positive results. This effort may involve a lengthy report or might simply be limited to formulating and outlining answers to a few fundamental questions. To take an extreme illustration, an individual who is considering feeding out 20 head of steers would not hire a consulting company to do a detailed study; on the other hand, a corporation that is considering investing several million dollars would be remiss if it did not give careful attention to all details that might prevent the venture from being a profitable one.

The outline of a cattle feasibility study, provided in Appendix 6.1, can be used as a table of contents for a written report or as a checklist if a more informal procedure is used. Many more aspects will be discovered as the analysis is carried out; thus, the outline should be considered only as a guide. Also, the amount of time spent on each section should be a function of its economic importance, as a constraint, or as information that can later be used by management rather that according to availability of information. In effect, the guide is a tool used to present information in an organized fashion rather than as an end in itself. Finally, although the outline provided in Appendix 6.1 uses cattle as an example, it can serve equally as well for other types of livestock. Further information can be found in Dyer and O'Mary (1972), Erickson and Phar (1970), and Richards and Karzan (1974).

Worldwide Experiences with Feedlots

The most important criteria for feeding grain to cattle is the beef-feed price ratio. In the United States and Canada it has typically been about 10-12 to 1 for maize--that is, the price per kg of the slaughter animal is about 10-12 times that of 1 kg of maize. The ratio has been about 7-8 to 1 in EEC countries, about 4-5 to 1 in Australia and Argentina, and 2-3 to 1 in many of the African countries. The ratio in the example problem presented in Table 6.9 is 8.83 to 1 on all costs ($1.43 cost per kg of live cattle divided by $0.1620 cost of maize). Examination of the ratios explains why feedlots have generally not been successful in countries outside the United States, except to a limited extent in the EEC (Schaefer-Kehnert, 1978).

A multitude of abandoned or little-used facilities attests to the failure of numerous attempts outside the United States and Canada to develop feedlots. In many countries such as Australia, New Zealand and the Central American countries, the decision to cease operation has usually been due to adverse world price movements--that is a failure to predict future trends for lots based on export markets.

The results from feedlot introduction have been mixed in Africa and are mostly from the late 1970s when world cattle prices were high. For example, Schaefer-Kehnert (1978) appeared rather pessimistic after reviewing the Kenyan experience. But Wardle (1979), in a study of small holder fattening in the Niger, was rather optimistic in reporting that a pilot scheme had been successful, although credit was found to be a constraint. Auriol reported in 1974 that much work needed to be done on evaluating the various types and combination of intensive feeding systems and that the main considerations, apart from economic relationships, were management and development of technical knowledge. Spring (1984) found that smallholder stall feeding in Malawi was successful but that incomes varied greatly depending on access to various management features and resources.

One way to overcome adverse grain cattle price ratios is to feed cattle by-products and crop residues (Jahnke, 1982). There are numerous reports on feeding straw, coffee pulp and animal wastes, but very fragmented work has been carried out on other products. Furthermore, there is a considerable range of research methodologies being used to identify which by-products are suitable for animal feed and to evaluate their availability, nutritional characteristics and potential for improvement through modification. It is apparent that despite important breakthroughs in improving the nutritive value of by-products, practical application of research results in existing animal feeding systems have been limited (Kiflewahid, Potts and Drysdale, 1983). The same is even true for rice straw (Doyle, Devendra and Pearce (1986).

It is becoming increasingly apparent that new feeding systems must be evaluated by means of careful comparison with methods currently in use. That means clear identification of clients and good ex-ante economic analyses to be sure the systems will be adopted (Myen, 1978). Perhaps most important is realization that effective by-product utilization does involve intensified production practices and that grain feeding, or silage use (which is a type of grain feeding), will often be required as a complementary feedstuff (Minish and Fox, 1982; Perry, 1980).

Cooperative Feedlots

Cattle feeding is at best a complex business due to the need for simultaneous coordination of a wide variety of interrelated parts. This task is even more complicated in developing countries, which generally lack an orientation toward efficiency as a key element in the livestock industry. Other constraints that may or may not be limiting in developing a feedlot project are input availability, an adequate and well-used grading system, market for the product, infrastructure in other sectors of the economy to provide support facilities, and technical know-how (Teele, 1984). Thus, even though the economics of confinement cattle feeding may appear viable, care must be taken to ensure commitment to, and an understanding of, all components involved, especially if the intention is to use the operation as a springboard for development activities (Squire and Creek, 1973; Squire, 1976).

One type of business organization for marketing livestock that has received considerable attention both in developed as well as developing countries is the cooperative. The idea is that a group of people, primary producers in this case, join together to participate in a business venture that they, as individuals, would not otherwise be able to accomplish. The cooperative can have a narrow focus such as cow/calf producers negotiating a contract to fatten dairy calves on milk replacer for sale as vealers, to very wide ranging objectives, such as assisting members who want to integrate forward into the growing phase, cattle feeding, or cattle slaughter. The cooperative might also market cattle and beef for its members (Ward, 1977). The activities can include identification, selection and negotiation of contracts with feedlots where a group of cattlemen retain ownership but not physical possession of cattle. The cooperative might even be formed to own a feedlot.

The benefits from cooperative ownership of agriculturally oriented facilities are potentially large and, as has been shown with grain elevators, feed supply stores and credit unions, can be very successful. Unfortunately, even though cooperative principles are sound in theory, application of them to ownership of a cattle feeding facility is difficult because of the commodity's nature. Cattle must be well taken care of and sold within a relatively restricted time period. For these and other reasons the manager must have considerable latitude to make decisions. This means that person must be well trained, reliable, honest and have an ability to get along with employees and management. Those criteria alone can become an overwhelming constraint.

Another problem with cooperatives is that cattle producers by nature are accustomed to managing their own affairs rather than working through a board of directors and participating in group decisionmaking of what is, essentially, a factory-type operation. A cooperative is a business organization that is harder to run than other forms of businesses enterprises, and its success depends in large part on maintaining sound business practices. For these reasons the cooperative approach generally has little applicability in the business of finishing cattle to slaughter weights.

Appendix 6.1 Typical outline of a cattle feeding
feasibility study

Chapter number	Item

Executive Summary

Table of Contents

List of Tables

List of Figures

I   Introduction

Problem statement

Objective and rationale

Area to be served

Purpose of the operation, such as provision of high quality meat versus cheap gain

II   Organization

Legal structure, such as cooperative, private corporation, partnership, government

Regulatory agency requirements, such as bonding, rate charges, accounting procedures

Management responsibilities

Level of management and supervision required

Potential of contracting management to a management company

III Area Cattle Feeding Structure

Number, location and volume of competing operations

History of competing operations; probable reasons for success or less than satisfactory operation

Reasons that other operations went defunct

Probable effect of new operation on existing competitors

Probably reaction from existing competitors

Intraregional advantage for the proposed location

Appendix 6.1 Typical outline of a cattle feeding
feasibility study (continued)

Chapter number	Item

IV Livestock Supply

Relevant supply area

Historical and projected trends in cattle supply

Structure of feeder-cattle-producing industry

Seasonality of supply

Type of cattle, age, classes and quality

Purchase arrangements and costs

Possibility of joint venture with feeder cattle producers and custom feeding

V Cattle Purchase, Assembly and Transportation

Use of own buyers, order buyers, auctions or other

Concentration or holding points

Direct delivery by producers

Transportation methods and regulations

Death loss, shrinkage, weather

Costs associated with cattle purchase and estimates of cost delivered to feeding areas

VI Feedstuffs and Feeding Practices

Feedstuffs source, availability, reliability, seasonality, and potential for expanded production

Prices of feedstuffs, fluctuations and reasons for the fluctuations

Competing use of feedstuffs

Government policies, restrictions on feedstuffs

Proposed rations and costs of them

Ration formulation and feeding method

Storage and purchase plans

Quality control plan for feedstuffs

Appendix 6.1. Typical outline of a cattle feeding
feasibility study (continued)

Chapter number	Item

VII Facility Requirement and Operating Plans

Drawing of facility

Description of process

Plans of structures

Adaptability of existing facilities and equipment;
alternative possibilities and reasons for one chosen

Rental versus owning facilities and equipment

VIII Marketing Fed Cattle

Projected demand for beef and fed cattle

Need to "create" a demand for meat from fed cattle

Price, cross and income elasticities of demand for the product and
similar products; analysis of their implications for the feeding
venture

Competition from other producers

Competition from imports; tariff and nontariff barriers

Potential for product differentiation

Current supply of cattle from competitors

Seasonality factors

Method and place of sale, such as auction markets, weighing at
feeding facility, or grade and yield at packing plant

Transportation availability and rates

Grades, grading standards, health standards

Profile of buyers and their market power

IX Summary of Feedlot Location Criteria

Origin of cattle

Location of feed source

Nearness to trading center

128

Appendix 6.1. Typical outline of a cattle feeding
                feasibility study (continued)

Chapter number	Item

Distance from feed processing plants if a complete feed is to be purchased

Nearness to slaughtering facilities

Proximity of marketing agencies, auctions and other livestock marketing channels

Adequacy of transportation facilities and/or a satisfactory highway network

Adequacy of service industries, such as labor force, veterinarians, mechanics, power, water

X **Economic Analysis**

Price trends of feeder and fed cattle including seasonality and week-to-week fluctuations

Determination of how prices are set and extent to which the proposed operation will affect price

Potential for price premiums due to cattle being feedlot finished

Estimate of seasonality effect on margins for cattle placed on feed at various times of the year

Budget for investment on total and per head basis for 3 sizes of operations (3 are needed to fit a curve for economies of size)

Calculation of investment cost per head by size of lot

Calculation of number of groups (lots) of cattle fed per year

Calculation of total cattle fed by lot size

Budget of operating expenses (total and per head fed) by size lot and two percentages of capacity (e.g., operating at 60 and 90 percent of capacity)

Calculation equipment replacement costs, annual basis

Calculation of total annual costs (investment, equipment replacement, and operating)

Calculation of total cost per head fed

Grouping cost per head fed to determine optimal size facility, given cattle supply conditions, funding availability, etc.

Appendix 6.1. Typical outline of a cattle feeding
feasibility study (continued)

Chapter number	Item

Estimates of breakeven volume by determining where total costs equal total revenue

Preparation of budget to determine breakeven cost, return on investment (same as budgets in Chapter 4)

Sensitivity of changes in cattle supply, meat demand, gains, prices and costs on profitability

XI Financial Analysis

Comparison of desired rate of return and ones calculated in budget

Source of funding for investment and operating costs

Debt repayment schedule

Monthly and annual cash flow

Determination of time required to begin making a profit, repay debts, etc., (cash flow)

Flexibility to be accorded to manager in financial decisionmaking

Factors that might cause actual cash flow and profits to differ from projections, such as delays in obtaining payments for sales, changes in interest rates, sickness of livestock, fluctuations in supply, and contingency plans for handling these factors

CHAPTER 7

GOAT AND SHEEP SYSTEMS AND ECONOMICS

About 95 percent of the world's goat inventory is found in developing countries. Of this population, 32 percent are in Africa, 57 percent in Asia and 6 percent in North and South America (Table 7.1). Goat numbers have fluctuated considerably during the past two decades; the direction of change has depended on the region. In tropical countries, goat populations have increased about 10 percent while in the Americas, USSR and Europe the populations have fallen. Numbers in Africa as a whole increased 12 percent, with most of the increase in temperate and subtropical areas. In the Asian countries quite different changes have also taken place, with goat inventory up substantially in the Philippines, but down in Indonesia and Thailand.

A very different pattern exists for sheep, as about 47 percent of world inventory is found in developed countries (Table 7.1). About 30 percent is located in Asia (compared with 57 percent of the world's goats) and 18 percent in Oceania (which has no appreciable goat numbers). The USSR and Europe each have 12 percent of sheep inventories but very small goat numbers.

The greatest proportion of goats are found in arid and semiarid areas, largely because goats are efficient users of the vegetation found in these areas (Gall, 1981). Milk accounts for about 58 percent of the total value of goat production, meat for about 36 percent, and skins and fiber for about 6 percent. Approximately 6 percent of the world's meat production is from goats. They also account for 6 percent of all hides and skins produced and 2 percent of milk production (Sands and McDowell, 1979). However, goat milk in both the developing and developed countries seldom enters commercial channels in adequate quantity or with sufficient regularity to be in competition with cow or buffalo milk (DeBoer, 1982).

One principal purpose of this Chapter is to explain how goat systems can be classified and to provide examples of the great diversity found in systems. A second purpose is to demonstrate the way economic analysis can be carried out on two very different problems with special attention

Table 7.1.   Inventory of goats and sheep by region, 1985

Region	Goats	Sheep
	- - - - - 1,000 - - - - -	
Africa	155,257	192,753
North & Central America	14,595	19,242
South America	19,819	102,283
Asia	250,522	311,155
Europe	12,555	133,331
Oceania	887	220,353
USSR	6,325	142,876
World	459,960	1,121,993
Developed	27,174	538,254
Developing	432,785	583,739
	- - - - - Percent - - - - -	
Africa	34	17
North & Central America	03	02
South America	04	9
Asia	54	28
Europe	03	12
Oceania	01	19
USSR	01	13
World	100	100
Developed	06	48
Developing	94	52

Source: FAO (1986).

given to the way cost and returns budgets can be developed when there are few purchased inputs. As with examples developed in earlier chapters, attention should be given to the procedures used in developing the economic analysis rather than the numbers even though the examples are based on actual situations.

Importance of Goats to Developing Countries

Goats are especially important in the developing world for a variety of reasons (Sandford, 1982 Devendra, 1987a). Some of these are

1. Subsistence consumption as food, materials for clothing, shelter, or other domestic equipment
2. Ritual, prestige and maintenance of social relations
3. Conjunctive herding with other animals
4. Maintenance of rangeland and range quality, especially for brush control

5. Generation of sustained cash income
6. High-yielding, relatively low-risk investment
7. Liquid asset for household necessities or for emergencies

An advantage of goats is they reproduce rather quickly (4.9-month gestation period), and thus optimal level stocking rates can be achieved more quickly than with cattle (which have a 9-month gestation period). This is especially important in arid areas in which prolonged dry spells result in major herd reductions. Also, while cattle are a high-investment resource, virtually no household finds a small number of sheep and goats impractical. Another advantage of goats is that their meat is often priced at a premium over beef. For example, in Kenya and Indonesia, goat meat is about 20 percent higher (DeBoer, 1981). A disadvantage is that the premium is often seasonal.

Goats seem to have a comparative advantage over dairy cows for home milk production by small landholders or nomads in some regions, especially arid and semiarid areas. Goats also have a comparative advantage for small farmers by virtue of the smaller investment and reduced risk. For example, it is not as catastrophic if one goat dies compared with one cow. But goats do not have a comparative advantage over dairy cattle in commercial production of fluid milk regardless of the type of ration fed.

Goats have a potential advantage over cattle in tropical climates in fecundity (annual number of young per breeding female). For example, under favorable conditions, goats have a parturition frequency of 8 months compared with 12 months for cattle, and there is a prolificacy of 1.5 kids per litter versus 1 calf. This results in a potential fecundity rate of 225 percent in goats versus 100 percent in cattle (Raun, 1982). However, goats have a higher rate of abortion, stillbirths and mortality of young animals. The net result is that in practice the fecundity rates for goats are 100-150 percent while for cattle the rates are 50-60 percent.

The fecundity rates are interesting and useful, but the more important aspect is the net value of the offtake of meat or milk per hectare, unit of labor, or other productivity and efficiency measures. Derivation of these calculations reveals that goats appear to have an advantage only in extensive or semi-intensive grazing systems in which inputs are held to a minimum. Also, marginal, (incremental), productivity increases in intensive goat production systems are seldom justified as purchased inputs and, labor management, along with problems of internal parasites, become proportionally greater in goats compared with other species. Also, as production intensity increases, nutritional requirements for goats appear to grow faster than for cattle (McDowell and Bove, 1977). The net result is that marginal returns to investment are not attractive nor competitive compared with other ruminants as system intensity grows (ILCA, 1980).

There are many production aspects in which goat productivity can be improved significantly (Sands and McDowell, 1978). One of these is improved disease resistance, especially to respiratory infections and gastrointestinal parasites. Another area is controlled mating (ILCA, 1979) to take advantage of feed availability on a seasonal basis and market price changes, and to reduce abortions and mortality. Milk production could be increased substantially, especially through selection (Sheton, 1978). Offtake can be improved through development of a systems approach to range or forage management and identification of the most economic role for goats within a whole livestock/farming enterprise (Cartwright and Blackburn, 1987).

Despite the drawbacks to goats as livestock husbandry intensifies, goats, along with sheep, are the most important and reliable animal protein sources in the Near and Middle East, North Africa and in the Mediterranean Basin (Fitzhugh, 1981). Goats are also very important in a high percentage of other Third World countries, especially the smaller ones (Oltenacu, Martinez, Glimp and Fitzhugh, 1976). Not only are a greater proportion of the world's goats and sheep found on farms of less than 5 hectares, but goats and sheep are also predominately found in the less industrialized countries. There is also a greater proportion of the total goat population found on small farms than is the case with sheep.

The integral role played by goats and sheep in world agriculture, and their increasing populations in low-income countries, suggest goats and sheep will increase despite the widespread misunderstanding of their legitimate role in the efficient use of natural resources. This is especially true in terms of the degradation and destruction of rangelands, pasturelands, and forests and in decertification of some areas. But at least one author (Raun, 1982) argues that the animals are not at fault; rather, the blame lies with poorly designed development programs, expansion of farming into unsuitable marginal land, and just plain poor management by livestock owners and people responsible for plan design and implementation.

Goats and sheep can effectively complement cattle if there is good management (Stoin, 1970). For example, Merrill (1980), describing a 21-year study in the goat and sheep area of Texas, shows that while cattle preferred grass, goats and sheep preferred forges. Goats consumed various kinds of browse after forage became short and ate grass readily even when dry. In general, simultaneous grazing by several classes of livestock resulted in a more uniform utilization of available forage and a reduction of poisonous plant problems than when only one kind of animal was grazed. Total economic returns were 25 percent higher for combination grazing than for only one species.

It has been observed that in low value range areas goats consume a larger number of plants, travel further in search of forages, and have a greater tendency to change diet with changing seasons than cattle do (Raun, 1982). Goats are also useful in clearing away understory of forested areas (Hansen and Child, 1980) and controlling scrub growth in pasture lands (Sheldon, 1980). Goats thus have the potential of being a substitute for manual, mechanical and chemical clearing operations.

Goat productivity can be increased by taking advantage of attributes inherent to the species. An essential prerequisite for high performance is maximum voluntary feed. In effect, rather than considering goats as a type of scavenger, attention is needed to effective physical, chemical or geological pretreatments of coarse crop residues to alleviate fibre digestion. Urea and ammonia treatment appear most promising (Devendra, 1987c). In addition, Devendra also points out that supplementation with energy, protein and minerals is necessary to correct nutrient limitation of which proteins are the most important. Economic use of preformed protein sources, non-protein nitrogen sources and urea-molasses block licks are all potentials. Nutritional strategies that are potentially important to sustain year-round feeding include wider and more intensive use of crop residues and agro-industrial by-products, expanded forage production on available land and strategic use of supplements. In effect, a critical point of departure in goat analysis work is focusing on nutritional requirements and from there carrying out economic analyses. An example about dairy goats in China is given in this chapter.

There are conflicting views about feed conversion efficiency of goats. Devendra (1975) feels that goats are more efficient than cattle in the conversion of animal feed into food for human consumption. In contrast, Van Soest (1980) arrives at the opposite conclusion. Settling the argument would be interesting from a physiological standpoint, but more important are economic considerations that can be approached using the theory and tools developed in earlier chapters. Such analyses fall into two categories, long term and short term. A short-term approach would answer the production economic questions of what to produce, how to produce and how much to produce in a static framework. There are many situations in developing countries where goats and/or sheep essentially must be part of the livestock mix regardless of the overall indications about optimal economic combinations. The theory behind this constraint will be developed by using the concepts of linear programming framework in a practical small-farm setting later in this book.

A more complex problem for the livestock system analyst is to determine long-term optimal livestock combinations--the "what to produce" problem in a dynamic model. One tool, dynamic linear programming, has been developed and applied to poly period problems such as this by Whitson (1976) for cattle operations. But the resources required to model these problems are prohibitive except for large regional projects, and, even then, noneconomic constraints will usually dictate that simpler-budgeting-oriented methods be used. But simulation models, discussed in a later chapter, can also provide usable, low-cost results to evaluate goat and sheep systems if a sophisticated modeling approach is desired (Blackburn, et. al., 1987).

We now turn to a method to classify goat systems and several case studies about them. Although this next section is specific to goats, the principles have relevance to analysis of sheep systems, which are covered in a following section.

Goat Systems

Goat systems, just like cattle systems, have evolved for the variety of reasons described in Chapter 1. There are a number of ways in which goat systems can be classified (e.f. Knipscheer, Hart and Baker, 1987). The approach taken in this book is to first divide them according to whether they are found in a modern or traditional setting. A final part of the process is to describe them according to zone and management characteristics typically found in each system (Table 7.2). As a rule, the more arid or semiarid an area, and the more traditional (less economically developed) it is, the more likelihood of finding goat operations. Furthermore, as the natural carrying capacity of a rangeland declines, the probability of finding nomadic and semi-nomadic systems increases. But as a country develops economically, and as population grows, pastoralists become more settled and goats become more of a business enterprise rather than constituting part of a subsistence operation.

The classifications in Table 7.2 are, of course, subject to wide variation given the range of conditions prevailing in each nation. For example, there are integrated village systems within the semi-intensive classification in which goats are part of small land holdings where producers emphasize tree or crop farming (Devendra, 1987b). As another example, there are some very rudimentary, extensive, meat-oriented goat operations using virtually no purchased inputs in the arid southwestern United States near urban centers, even though one would expect an intensive operation aimed at providing milk and using large amounts of purchased inputs because of the location and being in a "modern" society. In other words, because the purpose is analysis, Table 7.2 should be used as a standard or base and not as an absolute classification typical of the ones used in the biological sciences. Several case studies follow to exemplify the use of the classification scheme.

Iran: Case Study of an Extensive Traditional System

The area known as Khar Tauran in northeastern Iran is a region where goats predominate in the economy. Martin (1982) describes the zone as being primarily grazing land, but there are a number of settlements supported by a limited amount of irrigated agriculture based on the qanats, or underground channels, introduced in Iran during the first millennium B.C. Qanats carry ground water by gravity flow from an aquifer underlying relatively high ground to the surface at a point lower in a valley or plain where there is good soil. There are about 25,000 animals (primarily sheep) grazed year round in the area, and an additional 125,000 animals belonging to a linguistic group known as the Sangsari that venture into the area from the mountains east of Teheran to winter in the area. Families with small landholdings (10 hectares or less) co-mingle their animals for grazing in mixed village flocks that water at the individual villages. These families depend upon their goats primarily for family consumption of milk products and meat. Animals are rarely sold to meet immediate cash requirements; rather, cash is raised by shepherding for the Sangsari or larger local owners. Families with larger numbers (100 goats or more) try to obtain access to water and adjacent grazing areas away from the village for as long a time as possible.

Goat owners have participated in a market economy for a long time, despite the physical isolation from large urban centers, because the area lies adjacent to major caravan routes. Thus, it has not undergone the recent major shifts in pastoral production reported for some nomadic groups (Beck, 1980; Swift, 1979). Production of milk products is a major reason pastoralist keep goats, with many of the products being sold in the local economy from household to household, household to provincial market, and through middlemen to other households or provincial bazaars. Clarified butter is considered so important to pastoralists in Tauran that the output per animal is measured in the amount of clarified butter per animal. Output varies from 300-400 grams per animal in village flocks and up to 750-1,500 grams per animal in flocks away from villages. Other products are cheese and kasha, a sundried product rolled in a ball and made from dugh, or the liquid remaining after yogurt is churned and the butter removed. Kama, a product hardly known outside the area, is made by boiling dugh, pouring it into tanned skins, and then adding fresh dugh. The mixture thickens as it evaporates and is reconstituted for cooking.

Goats are sheared once a year in late spring or early summer. The hair is kept for ropemaking or sold to local part-time traders. The down, infrequently separated from the hair, is used for making hats, vests and scarves. Skins are usually saved for storage of fresh yogurt, cheese, Kameh and drained yogurt. Goat meat, priced the same as sheep meat, is marketed locally on an irregular basis. Sale of live animals is the major marketing decision due to distance from a large market. Martin's analysis agrees with other researchers (Low, 1980; Sandford, 1977; White, 1981) that the best understanding of sales decisions in a situation like this is not relating it to long-term price response, but rather to factors influencing marketing decisions. This is consistent with the holistic method that will be described in the chapter on small-farmer systems. Furthermore, small stock have to be sold virtually every year due to need for income.

Births (kidding) take place almost exclusively in late winter and early spring due to controlled breeding. Fattening thus begins in early spring. Most sales also occur in the spring and, to a reduced extent, in summer and early fall. Animals being fattened are grazed separately from the regular flocks and are the only animals with access to grazing on grain stubble and cotton stalks. These animals also receive supplementary barley, straw, and weeds pulled from fields as well as occasional alfalfa and sorghum if their owners raise it. Only limited fattening takes place due to a lack of irrigation water.

Table 7.2 Classification of goat systems for economic analysis

Item	Modern	Traditional
	Extensive	
Zone Characteristics	Low fertility land only suited to extensive livestock grazing.	
Management Characteristics	1. Moderate use of inputs. 2. Goats or sheep keep rather than cattle due to interest, soil characteristics or forage species. 3. No milking and little if any fattening.	1. Seminomadic or nomadic. 2. Very few, if any, purchased inputs. 3. Substance orientation 4. Little fattening, and milking only for family use. 5. Emphasis on quantity not quality.
	Semi-Intensive	
Zone Characteristics	Distance from urban areas not a factor.	Various alternative land uses.
Management Characteristics	1. Moderate use of purchased inputs. 2. Orientation to milking or fattening. 3. Possible integration with processing and/or delivery.	1. Minimal use of purchased inputs. 2. Used for meat, hair or wool production, and as a bank account. 3. Prestige and social acceptability important.

continued

Table 7.2   Classification of goat systems for economic analysis (continued)

Item	Modern	Traditional
	Intensive	
Zone Characteristics	Near urban areas or processing facilities or in areas with high yielding cropland.	
Management Characteristics	1. Extensive use of inputs. 2. Likelihood of milking orientation 3. Possible integration with processing and/or delivery.	1. Moderate use of inputs. 2. Likelihood of milking orientation 3. Probable delivery of milk.

Animals owned by large operators are often sent by them directly to Tehran while animals owned by smaller operators are mainly sold to Sangsari herders who migrate to their summer grazing area in the mountains near Teheran. The Sangsari move slowly and take advantage of new annual forage growth to fatten the animals as they move. Upon arrival the animals, which are mainly 2-year old males (kular), cull bucks, old does, and occasional pairs of does, are sold. The animals are largely transported by truck into the city. If animals are not sold in the spring, transportation to Tehran is a problem due to time (10-12 hours by truck) and uncertainty of obtaining transport.

Brazil: Case Study of a Semi-Intensive Traditional System

Another major small ruminant producing area is the semiarid zone of northeast Brazil, which had about 5.7 million sheep and 6.9 million goats in the mid-1970s. In contrast to the Iran study area, the northeastern Brazilian farms that specialize in sheep and/or goat production also have cattle, and cattle raising is the single most important economic activity. Cattle also receive preferential treatment over sheep and goats through allocation of the best grazing land, crop by-products and residues, and supplements. The average small ruminant flock of the 127 farmers in a 1980-81 sample was 211 head (Gutierrez and DeBoer, 1982). For those farms having goat herds (61 percent of the sample), average flock size was 116 head, while producers with sheep averaged 142 head. The predominant ownership pattern was mixed flocks. Specialized goat farms were rare (3 percent of the farms sampled).

In northeastern Brazil small ruminants are housed at night and allowed to graze extensively during the day. Animals are seldom herded, interior fencing is not a general practice, and straying to adjacent properties is common. Lack of dry matter production and deterioration of pasture quality during the dry season are the major constraints to improving carrying capacity and overall animal productivity. A wide range of supplements, including crop residues and commercial products, are used during critical dry periods. Supplementation priority is cattle first, then sick animals, followed by sheep. Goats are the lowest priority. Due to the feeding habits of goats and the availability of caatinga browse, they are better at surviving drought periods than are other animals and are considered as insurance against droughts. The main production strategy calls for sufficient flexibility to adjust animal numbers and herd composition quickly in response to drought or to periods of good rainfall.

Goats and sheep provide a regular supply of cheap, high quality protein to low income groups including the farmers themselves, agricultural laborers and urban workers. Sale of meat and by-products from small ruminants provides an important source of income even though the demand for small ruminant meat is such that retail meat prices for it are consistently less than with other red meats. Selling strategies favor sales directly at the farm due to more flexibility and better bargaining power, as well as avoiding the cost and inconvenience of taking animals to market.

West Africa: Case Study of a Semi-Intensive Traditional System

Another type of semi-intensive system is that found in West Africa, where small ruminants provide about 11 percent of the total meat supply in Ivory Coast and up to about 35 percent in Mali (Josserand and Ariza-Nino, 1982). International trade in small stock is important in this area because Mali, Upper Volta, and Niger have traditionally exported live sheep and goats (nearly 250,000 head annually) to all coastal countries in the region, from Senegal south to Nigeria and even northward to Algeria and Libya (Delgado, 1980).

Sheep and goats in West Africa are primarily kept for cash income, meat, milk, skins, and manure. The relative emphasis varies with the area considered and the type of local agricultural or stock-raising activities. For instance, cash income and meat are major objectives in all of West Africa, but milk yield of goats is given more importance in the Sahelian region than in coastal areas. In the Sahel goats are larger and often kept in herds of up to 200 animals. In coastal areas, where animals are smaller and milk yield is lower, emphasis shifts to meat production and droppings as a fertilizer source. In the Sahelian zone skins are used for tents, to carry water, and for making scabbards, wallets, and saddles, while sheep skins are used as prayer rugs. In the southern zones skins of small animals are consumed along with the meat or are grilled or pickled. Sheep and goats are kept throughout the rural areas and become progressively smaller in moving from the dry areas to the rain forest, probably because smaller animals are of breeds resistant to Trypanosomiasis. This same phenomena is found in cattle.

Most goats in Africa are kept under extensive conditions where they are herded on unfenced communal grazing areas (Wilson, 1987). It is unlikely that fencing can or will be used in the foreseeable future except in quite specialized cases such as near farmsteads because it is extremely expensive in relation to output value (Watts, 1982). This is especially true of most African situations as goats are mainly found in more arid areas where extensive amounts of fencing would be required. Regarding fencing, although there are some exceptions, the livestock analyst's problem in general is limited to ascertaining if fencing is feasible and then proceeding with evaluations of management practices or other analyses. The discussion on fencing is quite important as it drives home the point that livestock analysts, to be successful, must first ascertain the major constraint(s) in characterizing systems and, from there, begin to work out ideas for research on change. If, for example, fencing is not feasible, this has far flung implications for animal health programs, herding practices, use of children for labor, milk versus meat, type of breeds, etc.

In general, West African small ruminant owners live in rural areas and raise their animals on a continuing basis. In contrast, the small numbers of animals owned by urban dwellers are usually ones purchased for fattening and eating within a few weeks. Ownership of small ruminants is less concentrated on ethnic, social or economic grounds than on cattle. Thus, although it is not uncommon for a wife to have cattle of her own among the family herd, in the case of small ruminants personal ownership and individual decisions to buy or sell stock are the rule. Also, male and female adolescents frequently own small stock, dispose of them, and keep the sale proceeds.

Sheep and goats are mainly sold to raise cash, and, as such, the number marketed at any one time is small (Swift, 1981). They may be transported to a market place or sold in the family compound, but payment is virtually always in cash and is immediate. Females as well as males are traded, but mainly for reproduction purposes. Whereas cattle offal and other by-products are important sales items, this is not the case for goats, where edible offal only amounts to about 5 percent of liveweight and is sold along with the meat (Dahl and Hjort, 1976). The price of small ruminant meat is comparable to other common sources of animal protein.

The reason for keeping goats varies considerably between counties and across regions. In Africa there are many areas in which goats are not milked at all. Of the seven major breeds south of the Sahara, only the Nubian can be described as a milk breed (Watts, 1985, p. 35). In contrast, goats in Mediterranean countries make an important contribution to total milk supplies (Turner, 1978).

Economic Analysis of Milk Goats in China

A national inventory of about 63 million goats (at the end of 1985) makes the People's Republic of China (PRC) the second country in goats, closely following Brazil, which has about 82 million head. There is a great diversity in breed types and systems, which are largely dictated by geographic and climatic conditions (Jiang, An and Wang, 1987). Thus, for instance, 40 percent of China's cashmere production is found in Inner Mongolia, but that province only accounts for 15 percent of mohair production. Nevertheless, most of the cashmere and mohair production is found in the arid or extreme northern pastoral areas.

There are a wide variety of goat production systems in China. Milk goats are raised in three basic feeding systems: total confinement, partial confinement, and daytime grazing. Supplementation with concentrates is required in almost all cases. Provinces vary on policy, but some, such as Sichuan, have permitted producers supplying goat milk to state dairies to receive concentrates, with ratios varying from 1:3 to 1:10 depending on local policy.

There are no data on the number of milk breed versus other type of goats, but milk production of more than 400,000 tons annually indicates they are an important part of the goat subsector. If, for instance, milk production averaged 300 kg per doe, then there would be 1.33 million does. With replacements, dry does, bucks ,and so on, milk goat inventory could range from 1.8 to 2.0 million head, or about 3 percent of total goat inventory. Another factor pointing to the increasing importance of milk goats is that production has continually increased this decade.

An economic analysis is now carried out on two example dairy goat farms in Sichuan Province. One, called "confinement" has 5 purebred Saanen does that are permanently maintained in a confinement facility. The other, termed "grazing," is based on a herd of 10 Nubians that are grazed on communal pasture or alongside roads during the day and are confined at night.

Artificial insemination is used by both operations. Daily milk yield is 3.0 kg for the confinement farm and 2.5 kg on the grazing operation (Table 7.3). Total annual milk marketed is 3,250 kg in the former and 4,000 kg in the latter.

The confinement operation requires 11,915 Megacalories (Mcal) annually, while the grazing operation requires 25,342 Mcal. Concentrates, rated at 3.4 Mcal, are fed at a rate of 300 kg annually per doe on both operations. The remainder of Mcal requirements are met by feeding collected sweet potato vines (or equivalent forage) in the confinement operation and by grazing in the other one.

Net basic production costs (not including labor) are 1,274 yuan (Y) for the confinement operation and Y978 for the grazing one (Table 7.4). The latter is less expensive because no charge is made for grazing. Other costs are estimated at Y752 and Y3,568, respectively. Most of this cost is labor, which is charged at Y3.08 per day for general work and Y1.0 per day for herding. Total costs are Y1,766 and Y4,026 for the two units respectively.

There are two possible sales outlets, to the state and to private consumers. The price to the state is set at Y0.4 and to private consumers, Y0.8. In the case of state sales there is no delivery expense. Thus, net income above net basic production costs is Y26 and Y1,142 for the confinement and grazing operations respectively. However, there is a net loss of Y466 and Y2,426 on the two operations when all costs are considered. If labor is subtracted from all costs, then total net return to labor and management is Y242 and Y542 for the two systems.

Analysis of the private delivery system, in which an additional charge is made for delivery but where milk price is valued at Y0.8, reveals that net income above net basic production costs is Y1,586 and Y2,742 for the two operations. When all costs are considered, the net income is Y406 and Y-1,390.

Evaluation of the two systems shows that although an attempt can be made to reduce costs through grazing, the system does not provide as adequate a return to management as a confinement system does, even if only a very small charge is made for labor.

Table 7.3. Production parameters and price data for example dairy goat farms, Sichuan Province, 1984

Item	Units	System Confinement	Grazing
Breed	no	Saanens	Nubians
Milking does	no	5	10
Replacements	no	1	2
Male kids		Slaughtered at birth	Slaughtered at birth
Bucks	no	0	0
Total		6	12
		Production Parameters	
Annual milk yield per doe	kg	750	500
Lacation period	day	250	200
Daily milk yield	kg	3.0	2.5
Avg fat percentage	pct	4.0	5.0
Replacement rate	pct	20.0	20.0
Weaning weight	kg	7.0	7.0
ADG weaning to breeding	gms	100	50
ADG weaning to parturition	gms	100	100
Maintenance req. factor	pct	100	150
DE content of available DE (Mcal)	kg	2.88	2.40
Replacement doe age at			
Growing	mo	2-10	2-17
Pregnant	mo	10-15	17-22
Average body weight			
Lactating does	kg	55	50
Dry does	kg	60	55
Replacement, growing	lg	15.6	18.4
Replacement, pregnant	kg	38.9	37.3
Insemination	mhd	AI	AI
Milk per replacement	kg	100	100
Total annual milk production	kg	3,750	5,000
Total milk for replacement	kg	500	1,000
Total milk marketed	kg	3,250	4,000
Kiddings per doe per year	no	1	1
Fecundity	no	1.7	1.7

continued

Table 7.3. Production parameters and price data for example
dairy goat farms, Sichuan Province, 1984
(continued)

| | | System | |
Item	Units	Confinement	Grazing
		Feedstuffs	
Feeding method		Confinement only	Pasture, night con- finement
Feedstuffs		Cut grass, by-products	Native and semi- improved
Total herd feedstuff requirements per year			
Total protein	kg	476	979
Digestable energy	Mcal	11,916	25,342
Dry matter	kg	4,134	10,562
Pasture feedstuffs Requirements over confinement system			
Total protein	ratio	--	2.06
Digestible energy	ratio	--	2.13
Dry matter	ratio	--	2.55
Feedstuffs utilized per year			
Complete feed, 3.4 Mcal per kg (300 kg per doe)	kg	1,500	3,000
Cut grass, other greens	kg	19,474	0
		Labor input	
Forage collection	hr/day	2.0	0.0
Feeding & cleaning	hr/day	2.0	2.0
Milking	hr/day	1.0	2.0
Subtotal		5.0	4.0
Herding	hr/day	0.0	8.0
Milk delivery (if private)	hr/day	3.0	4.0

continued

Table 7.3. Production parameters and price data for example dairy goat farms, Sichuan Province, 1984 (continued)

Item	Units	System Confinement	Grazing
		Prices	
Labor			
General	yuan/day	3.08	3.08
Herding	yuan/day	--	1.00
Milk			
Private	yuan/kg	0.8	0.8
State	yuan/kg	0.4	0.4
Kids (female)	yuan/hd	30	30
Cull does	yuan/hd	50	50
Concentrate	yuan/kg	0.31	0.31
Forage	yuan/kg	0.04	--
Pasture	yuan	0.00	0.00

Source: Author estimates and calculations based on DeBoer (1984).
aThe exchange rate in 1987 was about Y3.7=$U.S. 1.00. Also, see footnotes to Table 7.4 for more information about parameters and price data.

The analysis is important for it shows how an economic analysis can be carried out when there are few apparent purchased inputs. The key is to adequately describe the systems and then to carefully develop the production parameters. Equally as important is recognition that magnitude rather than absolute size of results is important. In effect, the question usually revolves around a desire to determine the type of system or changes in the system, rather than just costs and return per se. Finally, the analysis shows how animal nutrition analyses can be related to goat economic studies.

Application to a Kenyan Sheep Enterprise

A review of the literature indicates that very few budgets have been published on small stock and that available budgets are difficult to interpret because the assumptions are not given. An important exception is an article by DeBoer, Job and Maundu (1982) that describes the budgeting of a number of goat and sheep systems in Kenya. One of these, a comparison of Red Maasai sheep with 3/4 Dorper by 1/4 Red

Table 7.4. Annual costs and returns for example dairy goat
farms in Sichuan Province, mid-1980s

Item	Confinement	Grazing
	- - - - Yuan - - - -	
Investment		
Land[a]	0	0
Building & equipment[b]	200	350
Goats[c]		
Mature does	400	800
Replacements	80	160
Total	680	1,310
Basic production costs		
Concentrate feed[d]	465	930
Forage[e]	779	0
Repairs & maintenance[f]	10	18
Veterinary[g]	5	10
Artificial innsemination[h]	5	10
Miscellaneous[i]	10	10
Total	1,274	978
Less salvage income		
Cull does[j]	50	100
Kids[k]	210	420
Total	260	520
Net basic production cost	1,014	458
Other costs		
Capital costs		
Land[l]	0	0
Building & equipment[m]	10	10
Goats[m]	24	48
Operating capital[n]	0	0
Subtotal	34	66
Ownership costs[o]		
Depreciation	10	18
Property tax	0	0
Insurance	0	0
Total	10	18
Labor, own		
General[p]	708	564
Herding[q]	0	2,920
Total	708	3,484
Total other costs	752	3,568
Total, all costs	1,766	4,026

continued

148

Table 7.4. Annual costs and returns for example dairy goat
farms in Sichuan Province, mid-1980s
(continued)

Item	Confinement	Grazing
	- - - - Yuan - - - -	
	Analysis, deliveries to state	
Income, milk	1,300	1,600
Net income, total		
Above net basic production costs	26	1,142
Above all costs	-466	-2,426
Net income per mature doe		
Above net basic production costs	5	114
Above all costs	-93	-243
Net income to labor & management		
Above all costs (subtract labor from costs)	242	542
	Analysis, private delivery	
Income, milk	2,600	3,200
Additional other costs (delivery labor)[r]	428	564
Total, all costs	2,194	4,590
Net income, total		
Above net basic production costs	1,586	2,742
Above all costs	406	-1,390
Net income, per mature doe		
Above net basic production costs	231	274
Above all costs	81	-139
Net income to labor & management		
Above all costs	2,722	2,658
Cost of milk production per kilo		
Above net basic production costs	.39	.24
Above all costs	.54	1.01

Source: Author estimates and calculations based on De Boer
(1984).

^aLand. No charge.
^bBuilding & equipment. 10 yuan per meter, 2 meter high house. 16m for 5 goats - 160 yuan + 40 yuan roof and so on - 200 yuan. The 10-goat facility is 30m - 300 yuan + 50 yuan other - 350 yuan.
^cGoats. Mature does and replacements at 80 yuan each.
^dConcentrate feed. Ration within the following (fresh basis)

Item	Percent ration	DE Mcal per kg	Total
Corn	50.00	3.50	175.00
Rapeseed cake	15.00	2.71	40.65
Rice bran	34.97	3.56	124.49
Salt	0.03	0	
Total			340.14
Adjust for percentage ration			3.40

Confinement operation
300 kg x 5 does - 1,500 kg x 3.40 - 5,100 Mcal. Then, 11,916 Mcal required minus 5,100 Mc - 6,816 Mcal divided by 0.35 (Mcal of sweet potato vines) = 19,474 kg vines (wet matter basis).

Feedstuff	Cost per kg	Ration	Cost per kg
	-Yuan-	-Percent-	-Yuan-
Corn	.38	.50	0.19
Rapeseed cake	.32	.15	0.05
Rice Bran	.20	.3497	0.07
Salt	.12	.03	0.00
Total		100	0.31

Concentrate cost is 1,500 kg (300 kg per doe) x 0.31 yuan - 465 yuan.
^eForage. Confinement 11,916 Mcal required minus 5,100 Mcal - 6,816 Mcal divided by 0.35 (Mcal of sweet potato vine) - 19,474 kg vines (wet matter basis at 0.04 per kg - 779 yuan).
  Grazing no charge assuming goats are on communal land.
^fRepairs and maintenance.   5 percent of buildings and equipment investment.
^gVeterinary. No charge for veterinarian. Materials 1 yuan per doe.
^hArtificial insemination.   1 yuan per doe.

Maasai cross, is now given to demonstrate both a method to budget small stock in developing countries as well as the application of partial budgeting for comparing two breeds.

The analysis compares two base flocks of 100 ewes set in a high-potential agroeconomic zone with more than 1,200, millimeters (mm) of rainfall (Table 7.5). Both systems follow the same management practices, such as culling, dipping and drenching. As a result, the cost structure for each is similar, 8,423 Kenyan shillings (Sh for the Red Maasai operation versus Sh 8,309 for the crossbred enterprise, Table 7.6). There are greater investment costs for building up the crossbred operation as Dorper rams must be purchased, but this is not budgeted as it is assumed that they can be obtained at a subsidized price from the government station. Also, numerous items such as investment and ownership costs that are included in total enterprise budgets need not be included as this is a partial analysis for comparative purposes. In effect, only costs that change must be budgeted. If the purpose were determination of cost per kilo, and/or indirect as well as direct costs, then the additional information would be required.

Results indicate that both total income (Sh 17,197) and net income (Sh 2,446) are considerably higher for the crossbred system than for the traditional one because of a higher output level (Table 7.6). One way in which the budgeted data can be manipulated to provide additional

---

iMiscellaneous. 2 yuan per doe.

jCull does. Replacement rate 20 pct. Value of 50 yuan per cull does.

kKids. 30 yuan each. 1.7 kids per doe x 5 = 8 kids minus 1 replacement = 7 kids for 5-doe herd and 14 for 10-doe herd.

lLand. No charge.

mBuilding & equipment, goats. 5 percent.

nOperating capital. No charge.

oOwnership costs. 20-year life on buildings and equipment for depreciation. No property taxes or insurance.

pLabor, general. Confinement operation, 0.63 (5 hrs per day) x 365 days = 230 days x 3.08 yuan per day = 708 yuan.
Grazing operation, 0.50 (4 hrs) x 365 = 183 days x 3.08 yuan = 564 yuan.

qLabor, herding. 8 hours per day @ 1 yuan = 8 yuan per day x 365 days = 2,920 yuan.

rAdditional costs. Confinement is 0.38 (3 hrs per day) x 365 = 139 x 3.08 yuan = 428 yuan.
Grazing is 0.50 (4 hours per day) x 365 = 183 hrs. x 3.08 yuan = 564 yuan.

Table 7.5. Production parameters and animal unit equivalents for two breeds of sheep, Red Maasai, and a cross breed 3/4 Dorper by 1/4 Maasai, Kenya, late 1970s

Item	Units	Red Maasai	Dorper X Red Maasai
Production parameters			
Ewes	no	100	100
Rams	no	2	2
Lambing	pct	102.0	92.0
Ewe replacement rate	pct	12.5	17.0
Ram replacement rate	pct	20.0	20.0
Dipping frequency	mo	1	1
Drenching frequency	mo	1	1
Cull ewe sale weight	kg	33	41
Lamb sale weight	kg	30	36
Mortality, matures	pct	5.0	5.0
Pre-weaning mortality	pct	9.0	12.0
Lamb sale age	mo	14	12
Age at joining	mo	16	12
Animal unit equivalents			
Ewes	au	20.0	23.0
Lambs	au	9.5	10.0
Ewe repl.& hoggets (lambs)	au	3.6	2.6
Rams	au	0.5	0.5
Total	au	33.6	36.1

Source: Adapted from De Boer, Job and Maundu (1982).

information is to place it on a per ewe and per AU basis as well as to convert the data from shillings to dollars. This transition is shown in Table 7.7.

There are several implicit assumptions to reach the higher output level with the crossbred sheep. The upgrading program, to have full impact, requires higher levels of management than are currently used on most Red Maasi sheep operations. For example, care must be taken in the crossbreeding program to avoid random rather than planned crossing, and it is necessary that adequate nutrition be provided to reach the heavier lamb sale weights. The question planners would then have to answer is whether government's cost in providing additional training as well as subsidizing the price of rams is excessive.

Table 7.6. Partial budget for comparing two breeds of sheep, Red Maasai, and a cross breed, 3/4 Dorper by 1/4 Red Maasai, Kenya, late 1970s

Dorper X Red Maasai

Additional Costs (sh)			Additional Income (sh)		
Drenching	14 sh/hd	2,772	Fat lambs	5.5 sh/kg	12,960
Dipping	6 sh/hd	1,188	Cull ewes	5.5 sh/kg	4,182
Replacement rams	sh	250	Cull rams	5.5 sh/kg	55
Labor (1 full-time person)	248 sh/hd	2,976	Total	sh	17,197
Minerals & salt	4.5 sh/hd	891			
Vaccinations	sh	232			
Total	sh	8,309			

Red Maasai

Reduced Income (sh)			Reduced Costs (sh)		
Fat lambs	5.5 sh/kg	11,550	Drenching	14 sh/hd	2,968
Cull ewes	5.5 sh/kg	3,267	Dipping	6 sh/hd	1,272
Cull rams	5.5 sh/kg	48	Replacement rams	sh	0
Total		14,865	Labor (1 full-time person)	248 sh/hd	2,976
			Minerals & salt	4.5 sh/hd	954
			Vaccinations	sh	253
			Total	sh	8,423

A. Total annual additional costs and reduced income (sh) 23,174

B. Total annual additional income and reduced costs (sh) 25,620

Net change in profit (B minus A). 2,446

Source: Adapted from De Boer, Job and Maundu (1982).

Table 7.7. Comparison of annual net returns for two breeds
of sheep on a total flock, per ewe and per AU
basis in shillings and U.S. dollars, Kenya, late
1970s

Item	Units	Maasai	Dorper X Red Maasai
Annual net returns			
Total flock	sh	6,442	8,888
Per ewe	sh	64.4	88.9
Per AU	sh	192	246
Total flock	U.S.$	805	1,111
Per ewe	U.S.$	8	11
Per AU	U.S.$	24	31

Source: Adapted from De Boer, Job and Maundu (1982).

CHAPTER 8

BUFFALO AND DAIRY CATTLE SYSTEMS
AND ECONOMIC ANALYSIS

Buffalo and dairy cattle are an extremely important part of many developing countries' economies. Buffalo provide draft power and often milk, and dairy cattle are a complement to many urban as well as rural households. The purpose of this chapter is to explain these type livestock in a systems framework and to extend the economic tools from earlier chapters to analytical problems found in them.

Draft Animals: A Vital Resource in Developing Countries

One of the more important uses of livestock in developing countries, and by smallholders in particular, is for draft power and various modes of transportation. Many species are included, with cattle and buffalo being the most widely utilized. Also important are horses, mules, donkeys, camels, elephants, llamas and yaks depending on geographic location and cultural background.

Draft animals are used for a wide variety of purposes. One of the more important tasks of the systems analyst is to identify the purposes as many of them, while rather subtle, are vital to a farmer's livelihood. For example, even though the use of water buffalo is widely recognized for ploughing, harrowing, and pumping water for irrigation, it is rather easy to overlook the equally important function water buffalo serve as a store of wealth and for thrashing grain. Mules and donkeys are used for transport, pack and field work in the Iberian Peninsula, parts of South America and Egypt, and other countries of the Near East (Smith, 1981). In desert and arid areas of the Near East, Africa and Asia, the camel is still considered an essential means of transport. In much of Southeast Asia, elephants are still used to a great extent for both work and transport.

Statistics are not available on the division of animals kept for meat versus draft purposes, but in Asia, for example, there are 4.2 million camels while there are 12.6 million head in Africa. In addition, there are about 375 million head of cattle in Asia and 174 million head in Africa, a high percentage of which are used for draft purposes in many countries.

Buffalo

Perhaps the most important of all draft animals is the buffalo, Bubalus bubalis, whose worldwide numbers are about one-ninth those of cattle. Buffalo constitute about one-third of total cattle, buffalo and camel inventory in Asia and about one-tenth of the total (of the three) on a worldwide basis. In Thailand there are about 50 percent more buffalo than cattle.

There are two general types of buffalo, the Swamp buffalo and the River buffalo. Swamp buffalo, usually slate gray, droopy necked, and oxlike with backswept horns, are found from the Philippines to as far west as India. They are primarily employed as a work animal and for meat, but never for milk production. River buffalo are found further west, from India to Egypt and up into Europe. They are usually black or dark gray with tightly coiled or drooping straight horns. In contrast to Swamp buffalo, which wallow in any water or mud puddle they can find, River buffalo prefer clean water. Also, River buffalo are a dairy type animal, accounting for almost 70 percent of the milk produced in India (National Research Council, 1981).

Water buffalo, as Swamp and River buffalo are collectively known, are bovine creatures, but are genetically further removed from cattle than are the North American bison. Thus, even though bison can be bred with cattle to produce hybrids (the F_1 males are sterile), there is no well-documented case of a mating between water buffalo and cattle that has produced progeny. However, embryos from water buffalo have been transferred to cattle and successfully carried to term.

Buffalo are among the gentlest of all farm animals. Also, and in contrast to generally held beliefs, buffalo are not just a tropical animal, as River buffalo are widely used in many temperate climate countries. Buffalo also produce a fine lean meat and rich milk that is higher in both butterfat and nonfat solids than milk from cattle is. Another benefit is that buffalo apparently have a more efficient digestive system than cattle because buffalo can extract nourishment from very poor and coarse forage. Finally, despite their usually spending considerable time in swamps, rivers, ponds and mudholes, foot diseases such as foot rot and foot abscesses are rare.

There are several drawbacks or limitations to use of buffalo regardless of their numerous outstanding qualities. They suffer greatly if forced to remain even for a few hours in direct sunlight without access to a wallow. Thus, they have little potential for arid areas. Also, time must be provided for wallowing, and buffalo cannot be driven long distances in the heat of the day. Sudden drops in temperature and chill winds can lead to pneumonia and death much more easily than is the case with cattle. Buffalo do not have the growth potential of cattle, espe-cially cattle crossbreds, and thus are less suitable for meat production per se. The gestation period of buffalo is about one month longer than that of cattle, and good management and nutrition are required to achieve high calf crops. They are as susceptible as most cattle are to diseases. Producers in Brazil, where large numbers of buffalo are now raised, know well that they are difficult to keep fenced in. Perhaps the main problem is that little research, especially at the farm level, has been carried out on buffalo.

Animal Versus Mechanical Power

One of the more difficult problems connected with draft animals is to set forth recommendations about the substitution of mechanical power in place of animals. The issue is especially complex because of the variety of factors involved. Some families may not want to shift from animals to mechanical power simply due to the importance of using animals as a store of wealth. Other families may not have the cash or economic wherewithal to make the transition even if they determine it to be in their interest. An example is a farmer who picks up milk from several neighbors and delivers it to a collection point using a cart drawn by oxen. Suppose that person could expand the service by shifting to a small tractor and wagon. An indirect benefit might be that sufficient time could be gained that would permit another activity, such as custom ploughing, to be initiated. But credit may be a constraint that cannot be overcome.

Risk is another often overlooked aspect when dealing with draft animals. Farmers in many irrigated areas of Asia, such as Thailand for instance, are rapidly adopting tractors as there is very little risk of a crop failure when irrigation is used that might prevent payments from being made on purchased equipment. Also, the fixed cost of the machinery can probably be spread over two crops rather than one (if that) in the rainfed areas. In contrast, draft animals are usually perceived to be a "free good" having been raised on the farm. Also, animals are virtually never purchased with time payments while credit is a ubiquitous feature of machinery purchase. Another dimension is that the closer one comes to a large municipal area, the more chance there if of finding tractors because the value of land is higher and this must be used on a more commercial basis. Furthermore, mechanics and spare parts are usually more readily available in irrigated as opposed to more remote areas. With rice land preparation (and most other crops) falling in a very short period, farmers have less risk in meeting planting deadlines when tractors are used rather than draft animals.

Draft animal research is deceiving for it appears quite simple, yet is very complex. For example, research in Africa has shown that a pair of oxen may also be associated with at least 3 cows, 1 bull, 2 female replacements, 1 bull replacement and 2 oxen replacements as well as the oxen themselves, a total of 11 head (Butterworth, 1982). The feed requirements, which are particularly high in the tropics where cattle performance is characterized by poor reproduction, slow growth and high mortality, are thus much higher than might at first be thought. Apart from feed as a drain on physical resources, labor use is much higher than might be suspected and in many cases may be more labor shifting than labor saving (Delgado and McIntire, 1982).

The evaluation of animal versus mechanical power is fraught with difficulties for there are very few purchased inputs employed in conjunction with draft animals. For example, improved harness may be provided by the state and as a result is a social cost rather than a cash outflow by the animal's owner. In some areas, animal dung is an important source of fuel, and the attachment of a family to the animals themselves versus a cold, hard piece of machinery cannot be overlooked. But although draft animals are used only for a restricted period of time on most farms, they require feed and care all year, and that can become quite costly.

The Peoples Republic of China (PRC) is one country where a rapid shift is taking place from animals to mechanical power in certain areas. Now that cropland has been decollectivized in the majority of China, many producers are finding that small, technologically adaptable tractors meet needs better than draft animals do. Rapid increases in farm income as a result of national price production and marketing policies are permitting farmers to purchase implements that just a few years ago were impossible to even dream about. It appears that the shift to mechanical power in the PRC, as in the rest of the world, now rests on size of holdings, cropping systems, cash and credit availability, formation of cooperative machinery use systems and availability of appropriate machinery.

The major conclusion one may draw after considering both direct and indirect factors related to draft power is that under some circumstances animals as draft power continue to be the best alternative and should be promoted, while under other conditions the use of machinery should receive research and extension priority. This is true not only among countries, but also within regions and even within a community. The situation of grain farming using oxen in the Sahel of Africa is entirely different than using buffalo for rice production in the Philippines. In effect, the systems and research are problem specific.

A great deal of work is needed on the nutrition, selection, breeding, health and other aspects of draft animals and their management (Goe, 1983). An example of the benefits that can be obtained with relatively little cost is incorporation of improved harness and more appropriate farming implements (Inns, 1980). Today, most of the yokes are handmade wooden affairs that are both inefficient and cruel (Ramaswamy, 1979). In Thailand, work carried out by the Agency for International Development (AID) showed that at least a 25 percent improvement in efficiency could be obtained by using a modified horse collar for buffalo rather than the traditional shoulder yolk, which hurts the shoulder due to small contact area and cuts off the animals's windpipe from the throat rope (Garner, 1980). Similar findings have been reported from various other countries

(Smith, 1981). Much work is being done on draft animals by the International Livestock Center for Africa (ILCA) at Addis Ababa, Ethiopia. But development of technologies still far exceeds adoption. In effect, there is a great gap between research and effective extension activities.

Quantitative analyses about interrelationships between work and milk production, and between nutrition and climate are only now beginning to appear (Reh and Horst, 1985). Furthermore, attention is needed on the tradeoff of draft versus losses in other goods and services from animals. For example, if there is only a limited amount of a particular by-product such as bran, it is questionable whether the optimal use is in feeding it to draft animals or to other ones such as a lactating cow or swine. It may be concluded that testing interventions related to draft animals must be carried out in full recognition of limited farmer resources and in the holistic framework of farming systems. In addition, research should cover one or more complete cropping seasons to reduce errors in measuring animal performance and to better understand interrelationships within households.

Cattle Versus Buffalo: An Example of Budgeting in Limited Input-Use Situations

One characteristic of small-producer livestock systems is limited input use, particularly of purchased inputs. In addition, the livestock products are frequently consumed by the owner's household or are traded or used as inputs in ancillary operations. For example, milk from yaks and goats is often made into cheese or fed to calves. The major use of buffalo may be to till cropland, while the primary purpose of cattle may be to pull carts rather than to produce beef. The net result is that few cells in the typical budgets presented in earlier chapters will be filled with cash expenses or income. We now examine the procedure to handle this situation.

Thailand, an Example

A typical budget developed by Tokrisna and Panayotou (1982) for both cattle and buffalo in Thailand and presented in Table 8.1 provides a reference point to solve the lack of cash inputs problem. The budget reflects traditional livestock production in Thailand where, as in many parts of the developing world, crop production is closely related and to a large extent is characterized by joint products and services. Buffalo and cattle are primarily raised for draft with meat production as a by-product at the end of their working life (about 10-12 years). Families typically own just 1-4 animals and use low-cost production technologies. Cattle and buffalo feed on harvested rice stubble during the dry season. Then, during the wet season when most fields are planted, cattle and buffalo are grazed on communal pastures, fallow fields, scrub forests, and rice paddy bunds. Cattle and buffalo will be fed rice straw and cut forage during much of the year as needed.

Table 8.1. Cost of raising cattle and buffalo, Thailand, 1981

	Buffalo			Cattle		
Item	Cash	Noncash	Total	Cash	Noncash	Total
	- - - - - - - - - - - - - - - U.S.$[a] - - - - - - - - - - - - - - -					
			Three year period			
Variable cost						
Labor						
Harvesting forage	--	43.60	43.60	--	28.90	28.90
Feeding	--	3.70	3.70	--	3.10	3.10
Tending	--	33.85	33.85	--	18.20	18.20
Watering	--	4.25	4.25	--	1.65	1.65
Subtotal	--	85.40	85.40	--	51.85	51.85
Materials						
Stock	--	37.50	37.50	--	25.00	25.00
Water & energy	--	11.20	11.20	--	7.35	7.35
Equipment	1.50	--	1.50	1.50	--	1.50
Medical care	--	1.80	1.80	--	1.80	1.80
Subtotal	1.50	50.50	52.00	1.50	34.15	35.65
Other						
Repairs	--	3.00	3.00	--	3.00	3.00
Opportunity cost of working capital	0.25	22.10	22.35	0.25	14.10	14.35
Subtotal	0.25	25.10	25.35	0.25	17.10	17.35
Subtotal, variable	1.75	161.00	162.75	1.75	103.10	104.85

continued

Table 8.1. Cost of raising cattle and buffalo, Thailand, 1981 (continued)

Item	Buffalo			Cattle		
	Cash	Noncash	Total	Cash	Noncash	Total
	- - - - - - - - - - - - - - - - - U.S.$[a] - - - - - - - - - - -					
Three year period						
Fixed cost						
Opportunity cost of land	--	28.40	28.40	--	16.45	16.45
Repair cost, stables	--	6.00	6.00	--	6.00	6.00
Depreciation, stables	--	6.00	6.00	--	4.50	4.50
Opportunity cost, capital	--	0.60	0.60	--	0.30	0.30
Subtotal, fixed	--	41.00	41.00	--	27.25	27.25
Total cost	1.75	202.00	203.75	1.75	130.35	132.10
Farmgate price	177.95	--	177.95	174.80	--	174.80
Profit	176.20	--	-25.80	173.05	--	42.70
Annual basis						
Total cost	0.58	67.33	67.92	0.58	43.45	44.03
Farmgate price	59.32	--	59.32	58.27	--	58.27
Profit above all cost	58.74	--	-8.60	57.69	--	14.24

[a]Exchange rate of 20 baht = U.S. $1.00.

Source: Tokrisna and Panayotou (1982) and based in budgets in NESDB, Agricultural Policy Planning, 1982-1986, Bangkok.

Analysis of the budget reveals that only about 1 percent of the costs associated with raising both buffalo and cattle are cash costs. There is a loss calculated of $25.80 during a 3-year period for raising buffalo compared with a positive $42.70 for cattle. Both operations assume the animals are raised for sale as meat, but the costs would be similar for animals raised as replacements for on-farm draft animals. The net income (profit) on an annual basis is a loss of $8.60 for buffalo but a gain of $14.24 per head of cattle.

The noncash expense items are calculated by imputing a value for each one based either on their market value or an opportunity cost. For example, feed can sometimes be sold to a neighbor (market value concept), or a farmer could work in alternative operations (opportunity cost). In fact, the budget reveals that about 40 percent of both operations' costs are for labor, very little of which may have direct alternative uses, such as working for cash wages, but to which an opportunity cost can be applied. The problem of imputing values for noncash costs is dealt with extensively by Brown (1979).

Pakistan, Another Example

There are various sources of data for budgets as well as outlines or formats for budgets. One approach is to collect the necessary quantity and price data and then prepare budgets for typical operations of the type being evaluated. These synthesized budgets are the type developed in Chapters 3 and 4. The advantage of synthesized budgets is relatively low preparation cost and direct applicability to particular problems at a specific place. Another way to prepare budgets is by carrying out a survey of producers to obtain costs and returns for a predetermined population. The difficulty is that even with a relatively homogeneous group, the average (mean) may be substantially different than the mode. Also, survey data are difficult to obtain as most producers dislike revealing total costs and returns even if they are known. Nevertheless, survey data can be useful, especially when the operations can be stratified (for example, according to size) or for year-to-year comparisons.

The problem of dealing with limited cash-use operations is now continued by examining results of survey data in a Pakistan study on two smallholder dairy systems. The budgets are presented in Table 8.2 in two formats, cash only and for all costs, that include an imputed market value for noncash items.

The budgets reveal that cash expenses are about 35-37 percent of total variable costs for each system; cash expenses are remarkably similar in that respect. In addition, there is a negative gross margin in both systems if only cash transactions are considered. But there is a substantial positive gross margin to the irrigated area system when imputed as well as cash costs and returns are both considered. The Semibarami area farmers about break even under this budgeting method.

It is important to recognize that a sizable proportion of total costs (both cash and noncash) are allocated to labor, 46 percent and 34 percent of the total for the two areas respectively. This is similar to the Thailand example where labor accounted for 40 percent of all costs. This situation is typical of most developing countries and reveals the need to give close attention to proper evaluation of the opportunity cost of labor. Finally, this Pakistani example, as well as the Thai example, demonstrates the importance of considering family consumption as a major part of income from smallholder livestock operations.

Dairy Systems and Economics

An introduction to dairy economics and systems was given in Chapter 3 as an example problem for budgeting and marginal analysis principles. Now the dairy industry is examined in more depth. A classification or categorization method is not developed, but analysts could easily develop one for their respective problem area by reviewing the methods for cattle and goats. But overall there are three main systems: intensive, semi-intensive, and subsistence. In general, intensive systems include relatively large commercial dairy farms with purebred cattle. Semi-intensive systems may have purebred or dual-purpose breeds. Subsistence-level operations will usually have 1-5 head of dual-purpose (milk/meat) or even draft-type animals. In some cases, as shown in a following example, the milk is sold. In other cases, it will be used directly by the family or fed to other farm animals.

Two example economic analyses are presented in this section, both of which are from the People's Republic of China. There are four major dairy production systems in China, one of which is pastoral based and provides milk either for family use or urban sale in grassland areas. The second encompasses very small private producers, usually with 1-4 cows in lactation. In this system, producers frequently herd their animals alongside roads or in communal grazing areas when forage is available. These producers also cut and carry forages and, during the winter, feed stored forages. A third, emerging type consists of small private producers with 5-25 cows in which the animals are maintained almost exclusively in barns and exercise yards. This type of enterprise will usually be part of an agricultural operation where the producer grows his or her own feedstuffs.

The last system, and by far the most important, is state farms and, to a lesser extent, collective (communal) farms. These production units are rather homogenous in most of China and center around substantial brick buildings with large upright silos. The barn will have a central walkway with cows tied on a gently sloping pad on each side.

The basic roughage in most areas is maize silage, but a wide variety of seasonal feedstuffs, such as surplus vegetables, sweet potato vines, fresh forage and hay along with crop residues like maize stover and rice straw, are commonly found. In addition, by-products such as wheat and rice bran, cotton seed meal, soybean meal and spent brewers grains are also commonly used.

Table 8.2.   Cost and returns per farm for two smallholder
             dairy systems in the Punjab, Pakistan, 1981

| Item | Irrigated area (near Lahore) 2.68 milk cows | | | |
	Cash gross margin	Per-cent	Imputed market value and direct costs[a]	Per-cent
	-U.S.$-		- U.S.$ -	
Variable cost				
Purchases	97.33	36	97.33	13
Feeding				
Green fodder	101.52	37	203.05	27
Dry fodder	31.71	11	57.92	8
Concentrate	27.05	10	27.05	4
Labor				
Family	--		341.01	46
Hired	--		--	
Grazing	--		--	
Breeding	--		--	
Animal health	2.63	1	2.63	0
Miscellaneous	13.32	5	13.32	2
Total	273.56	100	742.31	100
Cost per litre	0.17		0.46	
Income				
Milk	99.55	49	234.60	46
Butter	46.36	23	129.79	25
Ghee	9.27	4	27.81	5
Livestock sales	49.14	24	49.14	10
Farmyard manure	--	--	73.28	14
Total	204.32	100	514.62	100
Gross margin	-69.24		227.69	

continued

One problem is forage, which is variable in both quality and quantity. More serious is the widespread practice of simply feeding what is available without consideration of nutritive values or compatibility with other feedstuffs. In addition, forages found in communal areas or alongside roads are low quality and inferior to improved grass and legume varieties now readily available internationally. Variable feed quality also affects animal health. State and collective farms produce the bulk of their own feeds but sometimes contract with other townships (formerly called communes) or private producers primarily for roots, surplus vegetables, sweet potato vines and fresh grass.

Medium and large dairy producers usually mill their own maize, just adding a premix or other ingredients to make a compound feed. Sometimes these producers will purchase maize and mix with compound

Table 8.2. Cost and returns per farm for two smallholder
dairy systems in the Punjab, Pakistan, 1981
(continued)

| Item | Semibarami area (near Haveli) 1.8 milk cows | | | |
	Cash gross margin	Per-cent	Imputed market value and direct costs[a]	Per-cent
	-U.S.$-		- U.S.$ -	
Variable cost				
Purchases	36.19	19	36.19	7
Feeding				
Green fodder	75.10	39	227.56	41
Dry fodder	1.94	1	23.84	4
Concentrate	23.81	12	23.81	4
Labor				
Family	--		189.45	34
Hired	24.33	12	24.33	4
Grazing	23.93	12	23.93	4
Breeding	--		--	
Animal health	1.43	1	1.43	1
Miscellaneous	8.17	4	8.17	1
Total	194.90	100	558.71	100
Cost per litre	0.05		0.16	
Income				
Milk	38.71	22	114.30	20
Butter	--		--	
Ghee	126.53	72	379.60	67
Livestock sales	10.48	6	10.48	2
Farmyard manure	--		59.43	11
Total	175.72	100	563.81	100
Gross margin	-19.18		5.10	

[a]RS 10.50 - U.S. $1.00.
Source: Adapted from Rendall and Lockwood (1982).

feeds. Small producers will buy either a complete mix or a simple mixed
feed from one of the state-owned feed mills. In both cases, a production
subsidy is often provided in the form of reduced feed prices in proportion
to the amount of milk delivered to milk processing plants. In addition,
some provinces, such as Shanghai, have provided feedstuffs free of charge
on a proportion basis. A typical subsidy of this type has been 1 kg of
crop by-product for each 2.5 kg of milk supplied. The ratio for grain has
been 1 kg of grain for 4 kg of milk (Gartner and Krostitz, 1984). In
practice, state farms have first access to available subsidized feed

supplies. Collectives and small producers often must buy in the free market, which has considerable impact on their profits.

Specialized (very small) dairy producers often milk their cows four times per day and always at least three times. Collectives and state farms almost always milk three times. There is virtually no milking twice daily. In addition, there is little use made of milking machines except on very large operations. The reason for frequent milking and by hand is abundance of very cheap labor. In addition, milking machines are best justified when a completely modern milking parlor is introduced. But although introduction of such a system may be economically beneficial on some completely new operations, it would seem that a real push for such a system will only come when emphasis is placed on improving milk quality. At present, somatic cell counts are seldom made at milk plants.

Two cost and returns budgets for Chinese dairy operations are presented and analyzed in this section. Two sizes of northeastern China operations are evaluated, a small one with 4 mature cows and 1 replacement heifer, and a medium-size one with 200 mature cows and 60 replacement heifers. There are very few economies of size in dairy operations beyond 200 cows, so these budgets closely represent many of the large scale operations with 600-1,200 cows typical of state farms. This 200-head operation would be more representative of a collective dairy operation.

Production parameters and price data for the two example dairy farms are given in Table 8.3. Two situations are examined, one called "typical," which can be considered an average operation, while the other, designated "optimal," is a situation in which genetic stock and management are considerably above average.

The budget is divided into investment, basic production costs, other costs, income and cost per kilo (Table 8.4). Quantity, price and other information used to calculate costs are provided in footnotes to the table. Within the item "investment," there is no land charge because all land in China is state owned. However, land-use rights can be sold or traded, which means there is an opportunity cost for land that could become significant on farms near major cities. Assigning a value to this budget item would increase "other," that is, nonbasic, production costs.

Basal feed--the feed required to maintain a milking cow's condition--varies from 26 percent of basic production costs under the medium-size optimal operation to 41 percent for the small typical operator. Concentrate feed, which is charged at the existing government price--no production subsidy given on concentrate feed--varies from 32 to 55 percent of production costs depending on the type of operation. All feed costs vary from a low of 62 percent of production costs (typical medium operation) to a high of 83 percent for the optimum small-farm operation. Thus, as will be discussed in more detail later, most research and management attention on the cost side should focus on feed-use efficiency.

Table 8.3. Production parameters and price data for medium and small dairy cow farms in northeastern China, mid-1980s

Item	Units	Medium herds with purebreds Typical	Optimal	Small herds with lowgrade crossbreds Typical	Optimal
		Inventory			
Mature cows	hd	200	200	4	4
Replacement heifers	hd	60	60	1	1
Total animals	hd	260	260	5	5
		Production Parameter			
Calving rate	pct	75	92	66	86
Calving interval	mo	16	13	18	14
Death loss					
Weaned calves	pct	5	3	6	4
Cull cows	pct	1	1	1	1
First calving	mo	30	24	34	26
Replacement age	mo	72	66	84	72
Replacement rate	pct	29	29	24	26
Milk production per cow	kg	3,500	6,000	1,800	3,500
Sales weights					
Cows	kg	450	450	450	450
Bull veal calves	kg	40	40	40	40
Milkings per day	no	3	3	4	3
		Animals Marketed Annually			
Cull cows	hd	28	28	1	1
Veal calves					
Heifers	hd	14	59	-	-
Steers	hd	70	89	2	2
		Total Sale Weight			
Cull cows	kg	12,600	12,600	450	450
Calves	kg	3,360	3,560	80	80
Milk	kg	700,000	1,200,000	7,200	14,000

continued

The category "other costs" makes up about 20 percent of total expenses for both the medium and small operations (Table 8.5). These costs are separated from production costs because most are opportunity rather direct costs. For example, a charge is made to cover either or both these situations. A similar problem is the need to charge for depreciation--that is, for facilities and equipment that must be replaced at some point regardless of maintenance.

Table 8.3. Production parameters and price data for medium
and small dairy cow farms in northeastern China,
mid-1980s (continued)

Item	Units	Medium herds with purebreds		Small herds with lowgrade crossbreds	
		Typical	Optimal	Typical	Optimal
		Output Prices			
Cull cows	Yuan/kg	2.2	2.2	2.6	2.6
Calves	Yuan/kg	1	1	1	1
Milk	Yuan/kg	.55	.55	.55	.55
		Input Prices			
Complete feed	Yuan/kg	.26	.26	.26	.26
Corn silage	Yuan/kg	.03	.03	.03	.03
Hay	Yuan/kg	.09	.09	.09	.09
		Total Revenue			
Cull cows	Yuan	27,000	27,720	1,170	1,170
Calves	Yuan	3,360	3,560	80	80
Milk	Yuan	385,000	660,000	3,960	7,700
Total		415,360	691,280	5,210	8,950

Source: Collected by author on field trips.
3.7 Yuan = $1.00.

There is no charge to "own labor" in the medium-size operation because labor is hired. Independent producers do not charge themselves for labor but an opportunity cost must be calculated to avoid inflating the net income figures, especially when comparisons are made between operations. The appropriate wage rate varies considerably between producers, especially depending on their location. A charge of 80 yuan is used as a monthly base figure (3.7 yuan equaled $1.00 at the time this analysis was carried out).

Net income per mature cow is about Y500-700 on both small and medium-size operations and Y1,400-1,600 on optimal operations when only basic production costs are considered. It is still positive, Y200-1,000, when all costs are considered. These results are similar to reports made to the author on field surveys and also those reported by Gartner and Krostitz (1984) for Shanghai.

Cost per kilo of milk produced is less on small farms than on larger ones, but this advantage is offset by much lower income resulting from a lower genetic base. A key point is that planners must take net income, not just cost production into account when setting milk prices. Also, although it would initially appear that small producers are earning large profits, in reality that is not so when a charge is made for labor.

The net income figures will vary, of course, depending on location. For example, the price of maize increases from north to south due to

Table 8.4. Annual costs and returns for medium and small
           dairy cow farms in northeastern China, mid-1980s

Item	Medium (200 milking cow) herds with purbreds		Small herds with 4 milking cows	
	Typical	Optimal	Typical	Optimal
	- - - - - - - - Yuan - - - - - - - - - -			
Investment				
Land	0	0	0	0
Buildings & equipment[a]	400,000	400,000	6,000	6,000
Cattle[b]				
Cows	800,000	1,200,000	4,000	6,000
Heifers	180,000	300,000	800	1,000
Total	1,280,000	1,900,000	10,800	13,000
Basic production costs				
Hired labor[c]	72,000	72,000	0	0
Basal feed[d]	96,798	96,798	976	976
Concentrate feed[e]	97,614	163,114	936	1,945
Salt & minerals[f]	2,000	3,600	16	40
Repairs & maintenance[g]	20,000	20,000	300	300
Veterinary[h]	9,200	13,200	80	160
Artifical insemination[i]	300	736	2	7
Machinery[j]	5,000	5,000	45	50
Utilities[k]	2,000	2,000	0	0
Miscellaneous[l]	2,000	2,000	40	40
Total	306,912	378,448	2,395	3,518
Less salvage income				
Cull cows	27,720	27,720	1,170	1,170
Calves	3,360	3,560	80	80
Net basic production cost	276,552	347,168	1,145	2,268
Other costs				
Capital costs[m]				
Land	0	0	0	0
Buildings & equipment	20,000	20,000	300	300
Cattle	40,000	60,000	200	300
Operating capital	200	200	4	4
Subtotal	60,200	80,200	504	604

continued

Table 8.4. Annual costs and returns for medium and small
            dairy cow farms in northeastern China, mid-1980s
            (continued)

| Item | Medium (200 milking cow) herds with purbreds | | Small herds with 4 milking cows | |
	Typical	Optimal	Typical	Optimal
	- - - - - - - - - Yuan - - - - - - - - - - - -			
Ownership costs[n]				
Depreciation	13,333	13,333	200	200
Property tax	0	0	0	0
Insurance	0	0	0	0
Subtotal	13,333	13,333	200	200
Labor own[o]	0	0	1,200	1,200
Management, hired	1,800	1,800	0	0
Subtotal	1,800	1,800	1,200	1,200
Total, other costs	75,333	95,333	1,904	2,004
Total, all costs	351,885	442,501	3,049	4,272
Income[p]				
Milk	385,000	660,000	3,960	7,700
Net Income, total				
Above net basic				
production costs	108,448	312,832	2,815	5,432
Above all costs	33,835	217,499	911	3,428
Net income per mature cow				
Above net basic				
production costs	542	1,564	704	1,358
Above all costs	169	1,087	228	857
Net income own labor & management				
Above all costs				
Total	35,635	219,299	2,111	4,628
Per mature cow	178	1,096	528	1,157
Cost of milk production, per kilo				
Net basic				
production costs	0.40	0.29	0.17	0.17
All costs	0.50	0.37	0.43	0.31

continued

Table 8.4. Annual costs and returns for medium and small
            dairy cow farms in northeastern China, mid-1980s
            (continued)

Item	Medium (200 milking cow) herds with purbreds		Small herds with 4 milking cows	
	Typical	Optimal	Typical	Optimal
	- - - - - - - - - - U.S.$ - - - - - - -			
Net income per mature cow				
Above net basic				
production costs	146	423	190	367
Above all costs	46	294	62	232
Cost of milk production, per kilo				
Net basic production costs				
All costs				

Source: Author's estimates based on farm visits,
         publications and consultation with Chinese
         specialists.
         Exchange rate U.S.$1 = 3.7 yuan.

[a]Buildings & equipment. Includes hand milking facility with
all equipment, silo and exercise yard.
[b]Cows

Cattle	Medium		Small	
	Typical	Optimal	Typical	Optimal
	- - - - - - - - - - Yuan per head - - - - - - - - - - - -			
Cows	4,000	6,000	1,000	1,500
Heifers	3,000	5,000	800	1,000

[c]Labor. Ratio of 1 person per 5 cows in lactation. Includes
attendants for calves and replacements. Wage of Y150 per
month = Y1,800 annually, including bonuses.

continued

transportation cost. As an example, the free market price of maize was
about Y0.38-0.40 per kg in the northeast in August 1986. But it was
Y0.48 in Beijing, 0.50 in Jinan and 0.56 in Chengdu. Milk price was
reported to be Y0.55 in the northeast and Y0.50 in Beijing.
    All other factors being equal, Beijing producers would be at a
considerable disadvantage on both the cost and income side. But all
other factors are not equal as the state is heavily involved in providing
production subsidies. If, for example, the medium-size dairy producer

Footnotes Table 8.4. (continued)

^{d}Basic feed.

Feedstuff	Medium Quan-tity	Medium Price per kg	Medium Total cost	Small Quan-tity	Small Price per kg	Small Total cost
	-kg-	- - Y - -		-kg-	- - Y - -	
Corn silage	22	0.03	0.66	22	0.03	0.66
Hay	4	0.09	0.36	3	0.09	0.27
Total			1.02			0.93

		Total annual cost	
Milking cows			
Medium	365 days	74,460	--
Small	210 days	--	781
Replacements			
Medium	365 days	22,338	--
Small	210 days	195	
Total		96,798	976

Cattle are on pasture 155 days & fed stored feed 210 days.

Maintenance requirements are:

	Total requirements DCP	Total requirements TDN	Corn silage DCP	Corn silage TDN	Hay DCP	Hay TDN	Requirements met DCP	Requirements met TDN
	-GM-	-KG-	-GM-	-KG-	-GM-	-KG-	-GM-	-KG-
500 kg cow (Medium)	290	4.0	0.6	10.3	3.8	50.0	284	4.27
450 kg cow (Small)	270	3.7	0.6	10.3	3.8	50.0	246	3.77

^{e}Concentrate Feed. 0.5 kg feed per kg milk production

Cattle	Medium size Typical	Medium size Optimal	Small size Typical	Small size Optimal
	- - - - - Kilos feed per animal - - - - -			
Milking cows	1,750	3,000	900	1,750
Dry cows	120	180	--	120
Replacements	360	360	--	360
	- - - - Total kg feed annually - - - -			
Milking cows	350,000	600,000	3,600	7,000
Dry cows	3,840	5,760	--	120
Replacements	21,600	21,600	--	360
Total	375,440	627,360	3,600	7,480

Concentrate feed cost is 0.26 yuan per kg.

Footnotes Table 8.4. (continued)

f Salt and Minerals

Operation	Cost per cow unit (including replacements)
	- - - - - Yuan - - - - -
Medium	
Typical	10
Optimal	18
Small	
Typical	4
Optimal	10

g Repairs and Maintenance. 5 percent of buildings and equipment investment.

h Veterinary. No charge of veterinary services for small operations. Medium operation has one vet on staff @ 1,200 yuan per year.

Operation	Cost of supplies per cow
	- - - Yuan - - -
Medium	
Typical	40
Optimal	60
Small	
Typical	20
Optimal	40

i Artificial Insemination

Operation	Cost per conception	Calving rate	Conceptions per year
	- Yuan -	- Pct -	
Medium			
Typical	2.00	75	150
Optimal	4.00	92	184
Small			
Typical	0.50	66	2.64
Optimal	2.00	86	3.44

j Machinery. Allocation to tractor, wagon, silage loading, etc., including both owned and hired not covered in repairs and maintenance category.

k Utilities. Electricity, water. Medium size is 10 yuan per cow.

l Miscellaneous. 10 yuan per cow.

m Capital costs. 5 percent on buildings, equipment and cattle. Operating capital 1 yuan per cow.

174

were to receive a 25 percent reduction in the cost of concentrate feed, net income above all costs would increase 72 percent for the typical operation and 19 percent for the optimal producer (Table 8.6). In contrast, the impact is only a 26 and 14 percent increase for the two small-size producers respectively. The same pattern holds when only net basic production costs are considered.

_nOwnerships costs. 30-year life on buildings and equipment for depreciation. No property taxes or insurance.

_oOwn labor. Opportunity cost of people such as those owning dairy cattle is estimated to range between 70 and 100 yuan per month. An opportunity cost of 80 yuan is used. In summer tending animals while grazing can absorb one full person day, while milking and milk delivery to collection points absorb an additional one half day. In winter about one to 1.5 persons days is required depending on farm location. A weighted average of 1.25 person days is adapted.

Table 8.5. Percentage makeup of costs on example dairy cow
farms in northeastern China, mid-1980s

Item	Medium Typical	Medium Optimal	Small Typical	Small Optimal
	------- Percent[a] -------			
	Basic production costs only			
Hired labor	23	19	0	0
Basic feed	32	26	41	28
Concentrate feed	32	43	39	55
Salt & minerals	01	01	01	01
Repairs & maintenance	07	05	13	09
Veterinary	03	03	03	05
Artificial insemination	<1	<1	<1	<1
Machinery	02	01	02	01
Utilities	01	01	0	0
Miscellaneous	01	01	02	01
Total	100	100	100	100
	All costs			
Net basic production cost	79	79		
Other costs				
Capital costs	17	18		
Ownership costs	04	03		
Labor, own	0	0		
Management, hired	--	--		
Total	100	100		

Source: Table 8.4.
[a]Percentages many not add to 100 due to rounding errors.

Table 8.6. Impact of a 25 percent subsidy in feed concentrate on example dairy cow farms in northeastern China, mid-1980s

	Medium		Small	
	Typical	Optimal	Typical	Optimal
	- - - - - - - - Yuan - - - - - - - -			
Basic production costs				
Concentrate, no subsidy	97,614	163,114	936	1,945
25% subsidy on concentrate	24,404	40,779	234	486
Impact of 25 percent subsidy				
Net basic production cost	255,508	309,949	991	1,862
All cost	330,121	405,282	2,895	3,866
Cost of milk production per kilo				
Net basic production costs	.36	.26	.13	.13
All costs	.47	.33	.39	.27
Net income, total				
Above net basic production costs	132,852	353,611	3,049	5,918
Above all costs	58,239	258,278	1,145	3,914
Percent				
Net income, percent increase due to subsidy				
Above net basic production costs	23	13	8	9
Above all costs	72	19	26	14
Cost of milk production, percent decrease due to subsidy				
Net basic production costs	.10	.10	.24	.24
All costs	.06	.11	.09	.13

CHAPTER 9

CAPITAL BUDGETING:
THE TIME VALUE OF MONEY

The elements of budgeting and production theory were presented in Chapters 2 and 3 and budgets prepared for numerous types of livestock systems in the last five chapters. Various economic indicators, such as breakeven prices and production cost per kilo of animal produced, were calculated in many of the examples. In addition, the basics of linear programming were presented. In all cases the analyses were carried out in a static mode--for one point in time. But most investments, whether private or public, cover a number of years.

The added dimension of time is a focal point in this chapter. First, the concepts of compound interest and discounting are presented. Then, a classification of formulas used in time value of money problems are given along with examples. Following this, the partial budgeting technique is enriched by adding the time dimension to it. From there, capital budgeting as a private investment technique is explained. All of this sets the stage for the next chapter on project analysis or, as it has more traditionally been called, cost/benefit analysis.

The analysis of changes in an operation becomes somewhat more complicated when the costs and benefits occur in different time periods because a given sum of money is worth more today than at some point in the future. This is because money can be invested so that, at a future point, the investor would have the original sum plus interest. This concept is known as the time value of money.

As an example, assume that $1.00 is invested at 12 percent interest compounded annually for 10 years. The investor would receive $3.11 at the end of that period. Discounting is basically the reverse of compounding because the objective is to bring a future stream of benefits or costs back to the present. The parallel question from the compounding example is: what would $3.11 received 10 years from now be worth today given an opportunity cost of capital (the amount that could be received on money invested) of 12 percent. The answer is $1.00.

Formula Classification

There are a whole host of financial problems involving the time value of money. Some of the computations are time consuming and tedious if done by hand, but with the advent of low-cost pocket calculators and microcomputers containing programs for financial problems, the computations are now relatively easy. The more difficult part is to categorize the alternatives and to determine the appropriate formula for each problem.

A classification of the formulas is given in Figure 9.1 in an effort to overcome the confusion resulting from the myriad special cases. The classification is divided into two types of payments according to whether they are costs or income. A second breakdown is frequency of payments, either one time or recurring. Recurring payments are further subdivided into two categories, ones with the same amount at regular intervals and ones with varying amounts at irregular intervals. A final classification is the relationship of the conversion period to the payment period. The formulas, shown at the bottom of the figure, are divided into two types: those referring to future value (with a positive n) and the ones for present value computations (with a negative n). The positive n indicates a discount process. Discounting results in a smaller sum, while compounding results in a larger sum.

One-Time Cost or Income

The most basic problem in the time value of money, and the one used as the previous illustration is the formula for a one-time cost or income, that is, compound interest. The formula to calculate the future value of the $1.00 with an interest rate of 12 percent for 10 years is

$$A = P(1+i)^n$$

Where A = amount of future value

P = present value or principal

i = interest or discount rate per conversion period

n = number of conversion periods

As an example assume that a parcel of land is sold for $10,000 and the money invested for 10 years at 12 percent compounded annually (Table 9.1). The first step is to set up the problem

$$A = \$10,000 \ (1.12)^{10}$$

The factor for $(1 + i)^n$ is obtained by looking in Appendix 9.1 under the appropriate years (n) and interest rate (i). The result is a factor of 3.106; when multiplied by the $10,000 (it yields a value of $31,060, which would be recovered after 10 years.

CLASSIFICATION OF DATA AND FORMULAS USED

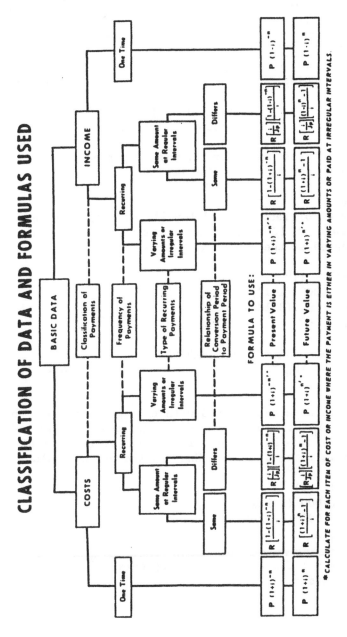

FIGURE 9.1 FORMULAS USED IN TIME VALUE OF MONEY PROBLEMS

SOURCE: ADAPTED FROM WALRATH (1977).

Table 9.1. Formulas and examples for problems with one-time
            cost or income

Example investment problem	Formula

1) Future amount of a present value
        (compounding)[a] $A = P(1 + i)^n$

Definition of symbols
                    A = amount of future value
        P = present value or principal
        i = interest rate per conversion period
        n = number of conversion periods

Example
    A parcel of land is sold for $10,000 and invested for
    10 years at 12 percent compounded annually. What is
    the value at the end of 10 years?

    $A = P(1 + i)^n$
    $A = \$10,000 \ (1.12)^{10}$
    $A = \$10,000 \ (3.106)$
    $A = \$31,060$

2) Present value of a future amount
        (discounting)[a] $P = A(1 + i)^{-n}$

Example
    You sell a parcel of land with the agreement made that
    the money will be invested at 12 percent compounded
    annually for 10 years. The value at the end of 10
    years is to be $10,000. What is the present value?

    $P = A(1 + i)^{-n}$
    $P = \$10,000 \ (1.12)^{-n}$
    $P = \$10,000 \ (0.322)$
    $P = \$3,220$

---

[a]See Appendix 9.1 for $(1 + i)^n$ and Appendix 9.2 for the
$(1 + i)^{-n}$ factors.

A very common application of this problem is opportunity-cost comparison problems. Opportunity cost is defined as the value of other opportunities given up in order to produce or consume another good. For example, if cattle prices drop it may be more profitable (or less costly) for a producer to simply sell part or all of his or her herd and invest the money in some other alternative investment, perhaps outside the livestock industry. The formula provides a quick way to calculate the alternative total return.

The present value of a future amount (discounting) is reverse of the preceding problem. In this case, the general formula is $P = A(1 + i)^{-n}$ with the factor found in Appendix 9.2. This factor is less than 1 because the problem is to drop down to a present value that, assuming a positive growth rate, would be smaller at the beginning of the time period. A sample problem is given in Table 9.1

Rates of Growth and Decline

Another use of the $(1 + i)^n$ formula is to ascertain annual growth rates, such as in a country's cattle population. If, for example, the inventory was 65,210,000 head in 1980 and 72,350,495 in 1985, what was the annual growth rate during this period? Division of the 1985 inventory by that of 1980 (72,350,495 ÷ 65,210,000) yields the coefficient 1.110. Because the two observations are five years apart, the factor can be found in the body of Appendix Table 9.1 along the row $n = 5$. The coefficient 1.110 falls between 1.104 (2 percent) and 1.131 (2.5 percent) meaning that the cattle inventory grew between 2.0 and 2.5 percent annually. If an exact percentage is desired, it can be found rather easily by the process of interpolation using ratios

$$\text{Interpolated segment of the } \underline{\text{growth}} \text{ rate} = \frac{\text{Absolute difference between lowest number and the coefficient}}{\text{Difference between highest number and the coefficient}} : \frac{X}{\text{Difference between the extreme growth rate points}}$$

$$= \frac{1.104 - 1.10}{1.131 - 1.10} : \frac{X}{2.5 - 2.0}$$

$$\frac{0.006}{0.027} : \frac{X}{0.5}$$

$$0.027X = 0.003$$

$$X = 0.11$$

The interpolated segment is <u>added</u> to the lowest growth rate, and therefore the growth rate is 2.0 + 0.11, or 2.11 percent. Thus, the cattle inventory grew at a compound rate of 2.1 percent annually from 1980 to 1985.

The $(1 + i)^{-n}$ table is used to calculate rates of decline. The procedure is similar to that given previously, except that the factors are found in Appendix 9.2. If, for example, the number of cattle for draft purposes was 5,478,000 head in 1965 and 3,152,000 in 1985, then division produces a coefficient of 0.576. The closest factors in the table under n = 20 years are 0.820 (1 percent) and 0.554 (3 percent). The interpolation procedure is

$$\text{Interpolated segment of the \underline{decline} rate} = \frac{0.554 - 0.576}{0.820 - 0.576} : \frac{X}{2.0}$$

$$= \frac{0.022}{0.244} : \frac{X}{2.0}$$

$$0.244X = 0.044$$

$$X = 0.18$$

The interpolated segment is <u>subtracted</u> from the highest growth rate (because this is a decline rather than growth problem), and therefore, the rate of decline is 3.0 - 0.18 = 2.82, or 2.8 percent annually.

Recurring Costs or Income

The next problem deals with costs and income that recur at regular intervals during the period being analyzed. These are called annuity problems. Again, values can be calculated either for the present or the future. First, let us calculate the future amount of a present value.

Assume a parcel of land is sold for $10,000, but rather than receive the whole sum today, an agreement is made to receive annual payments (at the beginning of each year) of $1,000. If that money were invested at 12 percent interest compounded annually, how much would that money be worth at the end of 10 years? The general formula is

$$S = R \frac{(1+i)^n - 1}{i}$$

Using the factor in Appendix 9.1, and solving the equation, leads to a value of $17,550. The computations are given in Table 9.2.

A parallel problem is to calculate the present value of an annuity, or a regular payment. If, for example, the land were sold for $10,000, but rather than receiving all the money at once, payments of $1,000 were received at the beginning of each year for 10 years, what is the effective sale price? Using Appendix 9.2 to locate the coefficient for $(1 + i)^{-n}$ it can be determined that the present value is $5,650 (Table 9.2).

Comparison of the Four Formulas

Let us have a partial summing up to compare results from the four formulas given in Tables 9.1 and 9.2. In all cases a decision was made to sell the land for $10,000. If all the money were received at the beginning of the time period, the value at the end of 10 years with a 12 percent annual interest rate and the money accumulating during the whole time period would be $31,058. If, on the other hand, annual payments of $1,000 per year were received and invested at 12 percent, the $10,000 sale price would be worth only $17,550 at the end of 10 years. In effect, there would be an opportunity loss of $13,508 from receiving annual payments rather than all of it initially. Another way to interpret this is that the annual payments strategy would only net 77 percent as much as receiving the lump sum at the beginning.

Another way to evaluate the sales alternative is to determine what the effective sale price, in today's money, would be from receiving annual payments of $1,000 versus one lump sum of $10,000 received at the end of 10 years. In this case the annual payment strategy yields a higher return ($5,560) than does the lump sum strategy ($3,220).

If the payments are received or made at more frequent intervals than annually, say quarterly or semiannually, the formula used becomes slightly more complicated, as can be seen in Figure 9.1. In this case the term i/jp is added where p = number of payments per conversion period and i/jp is a correction factor. Because of its limited use in livestock systems work, except for very specialized analyses, examples are not given.

The Internal Rate of Return

The time value of money concept and the basic formulas just presented can be used to determine the impact of different projects or possible business investments. In this section the use of discounting as an investment planning tool is discussed. The objective is to determine the rate of return on a possible investment by discounting the projected stream of benefits and costs back to the present in order to place them on a common denominator. This is called capital budgeting analysis. Let us take a short example.

Suppose that $6,000 is invested in a new well in an isolated area where previously there was little or no water for livestock. It is determined that as a result of the investment, additional annual benefits of

Table 9.2. Formulas and examples for problems with
recurring costs or income, regular intervals

Example investment problem	Formula

1) Future amount of a present value[a] $S = R \dfrac{(1 + i)^n - 1}{i}$

Definition of symbols
S = amount of future value
R = payment per conversion period
i = interest rate per conversion period
n = number of conversion periods

Example
You sell a parcel of land for $10,000, but rather than
receive the whole sum today, you agree to receive
annual payments (at the beginning of each year) of
$1,000. If that money were invested at 12 percent
interest compounded annually, how much would that
money be worth at the end of 10 years?

$$S = R \frac{(1 + i)^n - 1}{i} \qquad S = 1,000 \frac{2.106}{0.12}$$

$$S = 1,000 \frac{(1.12)^{10} - 1}{0.12} \qquad S = 1,000 \ (17.550)$$

$$S = 1,000 \frac{3.106}{0.12} \qquad S = 17,500$$

2) Present value of a future amount[a] $A = R \dfrac{1 - (1 + i)^{-n}}{i}$

Definition of symbols
A = present value
R = payment per conversion period
i = interest rate per conversion period
n = number of conversion periods

continued

Table 9.2. Formulas and examples for problems with recurring costs or income, regular intervals (continued)

Example investment problem	Formula

Example

You sell a parcel of land for $10,000, but rather than receive the whole sum today, you agree to receive annual payments (at the beginning of each year) of $1,000. If that money were invested at 12 percent interest compounded annually, how much would that money be worth at the end of 10 years?

$$A = R \; \frac{1 - (1+i)^{-n}}{i} \qquad\qquad A = 1,000 \; \frac{0.678}{0.12}$$

$$A = 1,000 \; \frac{1 - (1.12)^{-10}}{0.12} \qquad\qquad S = 1,000 \; (5.65)$$

$$A = 1,000 \; \frac{1 - 0.322}{0.12} \qquad\qquad A = \$5,650$$

[a]See Appendix 9.1 for $(1+i)^n$ and Appendix 9.2 for the $(1+i)^{-n}$ factors.

$2,000 per year will be gained. Let us also assume that no maintenance is required, the well has a useful life of five years, and the equipment has a salvage value of $500 at the end of year five. The question is whether to invest in the well and, if so, to determine the potential rate of return.

Computations presented in Table 9.3 show that the net benefits in year one are a negative $4,000 because an accounting convention in capital budgeting and project analysis is that all accounts are settled at year's end and that investments are assumed to be made on the first day of the year. The annual net benefits are $2,000 in all other years except for the last one, when the salvage value of $500 is also included.

The next step in calculating the internal rate of return involves choosing a discount rate that about equalizes discounted costs and benefits. In this case 20 percent was picked. For the first year the present value factor (given in Appendix 9.2) is 0.833. Multiplication of that coefficient times the negative $4,000 yields a negative $3,332. This multiplication procedure is carried out for each year, the column summed and a net present value calculated. It is a positive $105, which indicates that a higher discount rate must be tried as the objective is to bring the sum to zero.

The last step involves use of 25 percent, which yields a net present value of minus $201. Interpolating, as shown at the bottom of Table 9.3, gives an internal rate of return of 21.7 percent, which can be interpreted as meaning that the well project will yield an average net return of 21.7 percent each year during the projected useful life for all the money tied up in the project, including maintenance costs. If the opportunity exists to earn more than 21.7 percent in alternative projects, then those opportunities should be invested in rather than the well, providing there are no factors that preclude such investments.

A shortcut to the calculations is use of an annuity factor[1] for adjoining years in which the net benefits are the same. In the well problem the net benefits are $1,500 for years two-five so a discounting annuity factor, given in Appendix 9.3, can be used. The appropriate factor is obtained by subtracting the last year of the years previous to the series being computed. For example, the factor in Appendix 9.3 for year five at 20 percent is 2.991. The factor for the last year (year one) before the series begins is 0.833. The difference is 2.158, which is the present value factor for the computations.

The example demonstrates the ease with which discounting can be used. The internal rate of return is considered a superior method for it always gives correct ranking of investments. In contrast, the payback, benefit/cost ratio, net present worth, and rate of return methods can give incorrect rankings (Kay, 1981). Furthermore, the internal rate of return method also accounts for the time cost of money and vividly demonstrates the importance of including time and interest rates in investment decisionmaking. Inflation is incorporated as part of the internal rate of return, for inflation is implicit when the potential investor weighs opportunity cost of capital against the internal rate of return. High inflation rates tend to drive up interest rates so that the opportunity to invest at a higher rate in alternative projects is also much higher (Gittinger, 1972).

Evaluating Choice Between Investment Alternatives

It is often necessary to choose between different technologies or techniques that perform essentially the same function but have different time streams, investment periods, or maintenance costs. Examples are a wood versus a metal barn, an electric versus barbwire fence, or artificial insemination versus use of bulls. If it can be assumed that the benefits from two or more alternatives are equal, it is possible to evaluate the various systems strictly from the cost side using discounting procedures. There are four different possible situations or possibilities depending on the combination of time periods. These are

1. Coincident investment costs and retirements
2. Noncoincident investment costs and coincident retirements

Table 9.3. Calculations of pretax internal rate of return, example of investing in a well[a]

Year	Investment and maintenance	Annual benefits	Net benefits	20 Percent Present value factor[b]	20 Percent Present value	25 Percent Present value factor[b]	25 Percent Present value
			- - - - U.S.$ - - - -				
		Calculation of all years					
1	6,000	2,000	-4,000	.833	-3,332	.800	-3,200
2	500	2,000	1,500	.694	1,041	.640	960
3	500	2,000	1,500	.579	869	.512	768
4	500	2,000	1,500	.482	723	.410	615
5	500	2,000	1,500	.402	603	.328	492
5 (Salvage)		500	500	.402	201	.328	164
Net present value					+105		-201
		Use of annuity factors[c]					
1	6,000	2,000	-4,000	.833	-3,332	.800	-3,200
2-5	500	2,000	1,500	2.158	3,237	1.889	2,833
5	500		500	.402	201	.328	164
Net present value[d]					+106		-203

$$
IRR = \text{Lowest interest rate} + \text{Difference in interest rates} \times \left(\frac{\text{Present value, lowest interest rate}}{\text{Absolute present value, lowest interest rate} + \text{value, highest interest rate}} \right)
$$

$$
IRR = 20 + 5 \left(\frac{105}{105 + 201} \right) = 21.7
$$

b See Appendix 9.2 for discount factors.

c See Appendix 9.3 for annuity factors. For example, 2.158 = 2.991 - 0.833.

d The net present values are slightly different when annuity factors are used due to rounding errors from only using three place tables.

3. Coincident investment costs and noncoincident retirements
4. Noncoincident investment costs and noncoincident retirements

An empirical analysis of the different situations is now presented as a follow-up to the previous example on investing in a new well. The analysis is more detailed as there are two possibilities or choices called Plan I and Plan II. The problem is to determine the least-cost alternative. A discount rate of 12 percent is used for all examples.

Situation 1. Noncoincident Investment Costs and Coincident Retirements
A diagram of this problem is presented in the top part of Figure 9.2 while the computations are carried out in Table 9.4. The initial investments for both Plans I and II are made in year zero and retirements (the points at which the investment is either no longer functional or when it must be replaced) are also both in year ten (year shown in parentheses next to initial amount). The difference in assumptions is that Plan I has the highest investment ($10,000 versus $7,000 in Plan II) but lower operating costs ($500 annually versus $750). The number of years for which the computations are made, or year of the factor, are indicated by the figure in parentheses next to investment, salvage or operational cost. The appropriate appendix to obtain the factor is given in the footnotes.

The approach in this situation is, in summary, to obtain an annual worth of the investment using an annual annuity factor from a present amount (Appendix 9.4). Then, this annualized number is added to annual operating costs and the totals for the two plans ($2,270 versus $1,989) compared. Plan II is $281 less expensive per year; thus, this plan is the least-cost alternative and would be adopted.

Situation 2. Coincident Investment Costs and Coincident Retirements
The problem in this situation, graphically represented as the center diagram of Figure 9.2, with computations given in Table 9.4, is solved by calculating the present value (rather than annual value) of both investment and operating costs. The first step is to amortize the investment costs, which is done by determining the present worth (PW) of investment, salvage and operating costs. The term present worth is defined as the present value of a flow of money in future years. The approach here is to determine one lump sum at present whereas the previous example used annualized values.

Plan I has two parts, A and B. Part A is the initial investment cost including drilling, casing, motor, and so forth. Part B is a major modification in year five to permit expanded water discharge. These two costs are $10,000 and $5,000 respectively. In Plan II, the initial investment of $8,000 is higher than in Plan I, and furthermore, the substantial modification that takes place in year three rather than in year five is more expensive. The redeeming feature of Plan II is annual operating expenses are much lower ($1,620 versus $2,000) for part A of Plan I. A drawback is the annual expenses for the modification in Plan II are

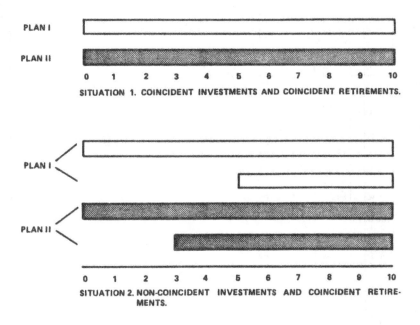

SITUATION 1. COINCIDENT INVESTMENTS AND COINCIDENT RETIREMENTS.

SITUATION 2. NON-COINCIDENT INVESTMENTS AND COINCIDENT RETIRE-
MENTS.

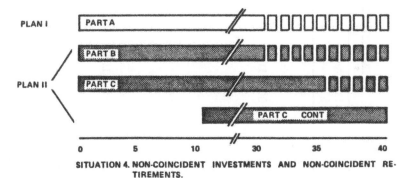

SITUATION 4. NON-COINCIDENT INVESTMENTS AND NON-COINCIDENT RE-
TIREMENTS.

FIGURE 9.2   THREE SITUATIONS FOR WELL EXAMPLE.

Table 9.4. Economic evaluation of two alternatives, situations 1 and 2, well installation example

Situation 1.
Noncoincident Investment Costs & Coincident Retirement

Item	Plan I Init. amt. & year - $ -	Plan I Factor[a]	Plan I Annual value - $ -	Plan II Init. amt. & year - $ -	Plan II Factor[a]	Plan II Annual value - $ -
Amortization of investment cost						
Part A						
First cost	10,000(10)	.177	1,770	7,000(10)	.177	1,239
Salvage						
Subtotal						
Part B						
First cost						
Salvage						
Subtotal						
Operating costs						
Part A						
Rearrangements						
Other operational costs	500(10)	--	500	750(10)	--	750
Part B						
Rearrangements						
Other operational costs						
Subtotal						
Total, annual value			2,270			1,989
Total, present worth						

continued

Table 9.4. (continued)

Situation 2.
Coincident Investment Costs & Coincident Retirements

Item	Plan I			Plan II		
	Init. amt. & year	Factor[b]	Annual value	Init. amt. & year	Factor[b]	Annual value
	- $ -		- $ -	- $ -		- $ -
Amortization of investment cost						
Part A						
First cost	10,000(1)	.893	8,930	8,000(1)	.893	7,144
Salvage	1,000(10)	.322	- 322	0 (10)	- -	0
Subtotal			8,608			7,144
Part B						
First cost	5,000(5)	.567	2,835	8,000(3)	.712	5,696
Salvage	50(10)	.322	- 16	0(10)	- -	0
Subtotal			2,891			5,696
Operating costs						
Part A						
Rearrangements	1,000(1)	.893	893	500(1)	.893	446
Other operational costs	2,000(10)	5.650	11,300	1,620(10)	5.650	9,153
Part B						
Rearrangements	300(5)	.567	170	500(3)	.712	356
Other operational costs	1,000(10-5)	2.613	2,613	1,620(10-3)	3.960	6,415
Subtota			114,976			16,370
Total, annual value			26,475			29,210
Total, present worth						

aAppendix 9.4.
bAppendix 9.2 and Appendix 9.3

higher ($1,620 versus $1,000). Because the operating costs recur on an annual basis for all years, the investment costs are brought to a present value by using the present value discount factors in Appendix 9.2 or the annuity factors in Appendix 9.3. The present worth of Plan I is $26,475, which is $2,735 less than the PW of Plan II. Thus, Plan I is the least cost alternative.

Situation 3. Coincident Investment Costs and Noncoincident Retirements

This situation is analyzed in the same manner as situation 2 in which there are noncoincident investment costs and coincident retirements. Thus, it is not illustrated graphically nor mathematically.

Situation 4. Noncoincident Investment Costs and Retirements

This situation is diagrammed in Figure 9.2 while the mathematical presentation is provided in Table 9.5. There are two plans, one with only one part called A while the other, Plan II has three parts, A, B, and C. The plans are placed on a comparable basis by first calculating values on an annual basis and then adjusting to a present-worth basis.

The investment costs are placed on an annual basis by multiplying the investment cost by the factor for year thirty (0.124) from Appendix 9.4. This is an annuity (annual basis) for a present amount. From this, the annualized salvage value is subtracted. The salvage value takes place in the last year of the project (year thirty), so it is multiplied by the factor (.004) of an annuity for a future amount. These factors are derived from the formula

$$\frac{i}{(1+i)^n - 1}$$

and are found in Appendix 9.5. Parts B and C are found in Appendix 9.5. Parts B and C are placed on an annual basis along with Part A by using the factor for year thirty, even though they take place at a later time (year forty), as shown in Figure 9.2.

The adjustment for noncoincident investment and retirement periods is carried out by placing the two different plans on a comparable time basis, forty years in this instance, because that is the maximum length of time covered by Plan II. In Plan I, the total annualized cost of $1,736 is multiplied by the factor (8.244) found in Appendix 9.3. The total, $14,312, is the present value of the investment and operating costs.

The adjustment from an annual basis to a present-worth basis is carried out for Plan II in the same way as in Plan I except that the present worth of an annuity factor must be calculated for parts B and C. In part B, for example, the factor is 3.605 in year five. Subtraction of this factor from 8.244 for year ten leaves 4.639. Each of the present values is summed to calculate the total present worth. Plan II is $2,023 lower, so it is the least-cost alternative.

Review of Appendixes

Five tables of coefficients, given in appendixes 9.1-9.5 are used in this chapter. It is useful to visualize these as groupings in order to clarify their use. The first one is compound interest, a relatively simple concept for calculating a future amount. Appendixes 2 and 3 are a group for discounting to the present--that is, to calculate one lump sum. Appendix 2 is for a single year while Appendix 3 is an annuity, (for multiple years). In Appendixes 2 and 3 future values are brought back to the present.

Appendixes 4 and 5 are used to place a single value on an annual basis. Appendix 4 is used to annualize a present cost while Appendix 5 is used to annualize a future income. The difference between them is important, although apparently small and is most easily understood by recalling the accounting convention that investments are assumed to be made on the first day of the first year. Thus, in Appendix 9.4, because the first year is included, an interest charge is made for that year when analyzing costs. But when the salvage value is received--when salvage value is an income rather than a cost--it is only necessary to consider the time period after the last day of the first year. For this reason the factor is 1.00 for all interest rates in year one.

NOTES

1. The term annuity can be slightly confusing. Note that the first use was synonymous with a recurring payment. An annuity factor, on the other hand, is the sum of a series of compound interest or discount factors.

Table 9.5. Economic evaluation of well installation example

Situation 4.

Noncoincident Investment Costs & Retirements

Item	Plan I			Plan II		
	Init. amt. & year	Factor[a]	Annual value	Init. amt. & year	Factor[a]	Annual value
	- $ -		- $ -	- $ -		- $ -
Amortization of investment cost						
Part A						
First cost	10,000(30)	.124[a]	1,240	4,000(30)	.124	196
Salvage	1,000(30)	.004[b]	- 4	400(30)	.004	- 2
Subtotal			1,236			494
Part B						
First cost				4,000(30)	.124	196
Salvage				400(30)	.004	- 2
Subtotal						494
Part C						
First cost				4,000(30)	.124	196
Salvage				400(30)	.004	- 2
Subtotal						494
Operating costs						
Part A	500		500	300		300
Part B				300		300
Part C				300		300

continued

Table 9.5. Economic evaluation of well installation example (continued)

Situation 4.

	Noncoincident Investment Costs & Retirements					
	Plan I			Plan II		
Item	Init. amt. & year	Factor[a]	Annual value	Init. amt. & year	Factor[a]	Annual value
	- $ -		- $ -	- $ -		- $ -
Total annual costs or present value						
Part A			1,736			794
Part B						794
Part C						794
Present worth						
Difference						
Total annual costs or present value						
Part A	1,736(40)	8.244[c]	14,312	794(40)[c]	8.244	6,546
Part B				794(40-5)[c]	4.639	3,683
Part C				794(40-10)[c]	2.594	2,060
Present worth			14,312			12,289
Difference						2,023

[a] Annual annuity from a present amount, Appendix 9.4.
[b] Annual annuity for a future amount, Appendix 9.5.
[c] Present worth of an annuity, Appendix 9.3.

Appendix 9.1 Compound interest, future value of 1 $(1 + i)^n$

n	.5%	1%	1.5%	2%	2.5%	3%	3.5%	4%	5%	6%
1	1.005	1.010	1.015	1.020	1.025	1.030	1.035	1.040	1.050	1.060
2	1.010	1.020	1.030	1.040	1.051	1.061	1.071	1.082	1.103	1.124
3	1.015	1.030	1.046	1.061	1.077	1.093	1.109	1.125	1.158	1.191
4	1.020	1.041	1.061	1.082	1.104	1.126	1.148	1.170	1.216	1.262
5	1.025	1.051	1.077	1.104	1.131	1.159	1.188	1.217	1.276	1.338
6	1.030	1.062	1.093	1.126	1.160	1.194	1.229	1.265	1.340	1.419
7	1.036	1.072	1.110	1.149	1.189	1.230	1.272	1.316	1.407	1.504
8	1.041	1.083	1.126	1.172	1.218	1.267	1.317	1.369	1.477	1.594
9	1.046	1.094	1.143	1.195	1.249	1.305	1.363	1.423	1.551	1.690
10	1.051	1.105	1.161	1.219	1.280	1.344	1.411	1.480	1.629	1.791
11	1.056	1.116	1.178	1.243	1.312	1.384	1.460	1.539	1.710	1.898
12	1.062	1.127	1.196	1.268	1.345	1.426	1.511	1.601	1.796	2.012
13	1.067	1.138	1.214	1.294	1.379	1.469	1.564	1.665	1.886	2.133
14	1.072	1.149	1.232	1.319	1.413	1.513	1.619	1.732	1.980	2.261
15	1.078	1.161	1.250	1.346	1.448	1.558	1.675	1.801	2.079	2.397
16	1.083	1.173	1.269	1.373	1.485	1.605	1.734	1.873	2.183	2.540
17	1.088	1.184	1.288	1.400	1.522	1.653	1.795	1.948	2.292	2.693
18	1.094	1.196	1.307	1.428	1.560	1.702	1.857	2.026	2.407	2.854
19	1.099	1.208	1.327	1.457	1.599	1.754	1.923	2.107	2.527	3.026
20	1.105	1.220	1.347	1.486	1.639	1.806	1.990	2.191	2.653	3.207
25	1.133	1.282	1.451	1.641	1.854	2.094	2.363	2.666	3.386	4.292
30	1.161	1.348	1.563	1.811	2.098	2.427	2.807	3.243	4.322	5.743
40	1.221	1.489	1.814	2.208	2.685	3.262	3.959	4.801	7.040	10.286
50	1.283	1.645	2.105	2.692	3.437	4.384	5.585	7.107	11.467	18.420

n	7%	8%	9%	10%	12%	14%	16%	20%	25%
1	1.070	1.080	1.090	1.100	1.120	1.140	1.160	1.200	1.250
2	1.145	1.166	1.188	1.210	1.254	1.300	1.346	1.440	1.562
3	1.225	1.260	1.295	1.331	1.405	1.482	1.561	1.728	1.953
4	1.311	1.360	1.412	1.464	1.574	1.689	1.811	2.074	2.441
5	1.403	1.469	1.539	1.611	1.762	1.925	2.100	2.488	3.052
6	1.501	1.587	1.677	1.772	1.974	2.195	2.436	2.986	3.815
7	1.606	1.714	1.828	1.949	2.211	2.502	2.826	3.583	4.768
8	1.718	1.851	1.993	2.144	2.476	2.853	3.278	4.300	5.960
9	1.838	1.999	2.172	2.358	2.773	3.252	3.803	5.160	7.451
10	1.967	2.159	2.367	2.594	3.106	3.707	4.411	6.192	9.313
11	2.105	2.332	2.580	2.853	3.479	4.226	5.117	7.430	11.462
12	2.252	2.518	2.813	3.138	3.896	4.818	5.936	8.916	14.552
13	2.410	2.720	3.066	3.452	4.363	5.492	6.886	10.699	18.190
14	2.579	2.937	3.342	3.797	4.887	6.261	7.988	12.839	22.737
15	2.759	3.172	3.642	4.177	5.474	7.138	9.266	15.407	28.421
16	2.952	3.426	3.970	4.595	6.130	8.137	10.748	18.488	35.527
17	3.159	3.700	4.328	5.054	6.866	9.276	12.468	22.186	44.409
18	3.380	3.996	4.717	5.560	7.690	10.575	14.463	26.623	55.511
19	3.617	4.316	5.142	6.116	8.613	12.056	16.777	31.948	69.389
20	3.870	4.661	5.604	6.727	9.646	13.743	19.461	38.338	86.736
25	5.427	6.848	8.623	10.835	17.000	26.462	40.874	95.396	264.698
30	7.612	10.063	13.268	17.449	29.960	50.950	85.850	237.376	807.794
40	14.974	21.725	31.409	45.259	93.051	188.883	378.721	1469.772	7523.164
50	29.457	46.902	74.358	117.390	289.002	700.233	1670.704	9100.438	70064.923

Appendix 9.2 Discounting, present value of $1 (1 + i)^{-n}$

n	.5%	1%	1.5%	2%	2.5%	3%	3.5%	4%	5%	6%
1	.995	.990	.985	.980	.976	.971	.966	.962	.952	.943
2	.990	.980	.971	.961	.952	.943	.934	.925	.907	.890
3	.985	.971	.956	.942	.929	.915	.902	.889	.864	.840
4	.980	.961	.942	.924	.906	.888	.871	.855	.823	.792
5	.975	.951	.928	.906	.884	.863	.842	.822	.784	.747
6	.971	.942	.915	.888	.862	.837	.814	.790	.746	.705
7	.966	.933	.901	.871	.841	.813	.786	.760	.711	.665
8	.961	.923	.888	.853	.821	.789	.759	.731	.677	.627
9	.956	.914	.875	.837	.801	.766	.734	.703	.645	.592
10	.951	.905	.862	.820	.781	.744	.709	.676	.614	.558
11	.947	.896	.849	.804	.762	.722	.685	.650	.585	.527
12	.942	.887	.836	.788	.744	.701	.662	.625	.557	.497
13	.937	.879	.824	.773	.725	.681	.639	.601	.530	.469
14	.933	.870	.812	.758	.708	.661	.618	.577	.505	.442
15	.928	.861	.800	.743	.690	.642	.597	.555	.481	.417
16	.923	.853	.788	.728	.674	.623	.577	.534	.458	.394
17	.919	.844	.776	.714	.657	.605	.557	.513	.436	.371
18	.914	.836	.765	.700	.641	.587	.538	.494	.416	.350
19	.910	.828	.754	.686	.626	.570	.520	.475	.396	.331
20	.905	.820	.742	.673	.610	.554	.503	.456	.377	.312
25	.833	.780	.689	.610	.539	.478	.423	.375	.295	.233
30	.861	.742	.640	.552	.477	.412	.356	.308	.231	.174
40	.819	.672	.551	.453	.372	.307	.253	.208	.142	.097
50	.779	.608	.457	.372	.291	.228	.179	.141	.087	.054

n	7%	8%	9%	10%	12%	14%	16%	20%	25%
1	.935	.926	.917	.909	.893	.877	.862	.833	.800
2	.873	.857	.842	.826	.797	.769	.743	.694	.640
3	.816	.794	.772	.751	.712	.675	.641	.579	.512
4	.763	.735	.708	.683	.636	.592	.552	.482	.410
5	.713	.681	.650	.621	.567	.519	.476	.402	.328
6	.666	.630	.596	.564	.507	.456	.410	.335	.262
7	.623	.583	.547	.513	.452	.400	.354	.279	.210
8	.582	.540	.502	.467	.404	.351	.305	.233	.168
9	.544	.500	.460	.424	.361	.308	.263	.194	.134
10	.508	.463	.422	.386	.322	.270	.227	.162	.107
11	.475	.429	.388	.350	.287	.237	.195	.135	.086
12	.444	.397	.356	.319	.257	.208	.168	.112	.069
13	.415	.368	.326	.290	.229	.182	.145	.093	.055
14	.388	.340	.299	.263	.205	.160	.125	.078	.044
15	.362	.315	.275	.239	.183	.140	.108	.065	.035
16	.339	.292	.252	.218	.163	.123	.093	.054	.028
17	.317	.270	.231	.198	.146	.108	.080	.045	.023
18	.296	.250	.212	.180	.130	.095	.069	.038	.018
19	.277	.232	.194	.164	.116	.083	.060	.031	.014
20	.258	.215	.178	.149	.104	.073	.051	.026	.012
25	.184	.146	.116	.092	.059	.038	.024	.010	.004
30	.131	.099	.075	.057	.033	.020	.012	.004	.001
40	.067	.046	.032	.022	.011	.005	.003	.001	.000
50	.034	.021	.013	.009	.004	.001	.001	.000	.000

Appendix 9.3 Discounting annuity factors $\dfrac{1 - (1 + i)^{-n}}{i}$

n	.5%	1%	1.5%	2%	2.5%	3%	3.5%	4%	5%	6%
1	.995	.990	.985	.980	.976	.971	.966	.962	.952	.943
2	1.985	1.970	1.956	1.942	1.927	1.913	1.900	1.886	1.859	1.833
3	2.970	2.941	2.912	2.884	2.856	2.829	2.802	2.775	2.723	2.673
4	3.950	3.902	3.854	3.808	3.762	3.717	3.673	3.630	3.546	3.465
5	4.926	4.853	4.783	4.713	4.646	4.580	4.515	4.452	4.329	4.212
6	5.896	5.795	5.697	5.601	5.508	5.417	5.329	5.242	5.076	4.917
7	6.862	6.728	6.598	6.472	6.349	6.230	6.115	6.002	5.786	5.582
8	7.823	7.652	7.486	7.325	7.170	7.020	6.874	6.733	6.463	6.210
9	8.779	8.566	8.361	8.162	7.971	7.786	7.608	7.435	7.108	6.802
10	9.730	9.471	9.222	8.893	8.752	8.530	8.317	8.111	7.722	7.360
11	10.677	10.368	10.071	9.787	9.514	9.253	9.002	8.760	8.306	7.887
12	11.619	11.255	10.908	10.575	10.258	9.954	9.663	9.385	8.863	8.384
13	12.556	12.134	11.732	11.348	10.983	10.635	10.303	9.986	9.394	8.853
14	13.489	13.004	12.543	12.106	11.691	11.296	10.921	10.563	9.899	9.295
15	14.417	13.865	13.343	12.849	12.381	11.938	11.517	11.118	10.380	9.712
16	15.340	14.718	14.131	13.578	13.055	12.561	12.094	11.652	10.838	10.106
17	16.259	15.562	14.908	14.292	13.712	13.166	12.651	12.166	11.274	10.477
18	17.173	16.398	15.673	14.992	14.353	13.753	13.190	12.659	11.690	10.828
19	18.082	17.226	16.426	15.678	14.979	14.324	13.710	13.134	12.085	11.158
20	18.987	18.046	17.169	16.351	15.589	14.877	14.212	13.590	12.462	11.470
25	23.446	22.023	20.720	19.523	18.424	17.413	16.482	15.622	14.094	12.783
30	27.794	25.808	24.016	22.397	20.930	19.600	18.392	17.292	15.373	13.765
40	36.172	32.838	29.916	27.355	25.103	23.115	21.355	19.793	17.159	15.046
50	44.143	39.196	35.000	31.424	28.362	25.730	23.456	21.482	18.256	15.762

n	7%	8%	9%	10%	12%	14%	16%	20%	25%
1	.935	.926	.917	.909	.893	.877	.862	.833	.800
2	1.808	1.783	1.759	1.736	1.690	1.647	1.605	1.528	1.440
3	2.624	2.577	2.531	2.487	2.402	2.322	2.245	2.106	1.952
4	3.387	3.312	3.240	3.170	3.037	2.914	2.798	2.589	2.362
5	4.100	3.993	3.890	3.791	3.605	3.433	3.274	2.991	2.689
6	4.767	4.623	4.486	4.355	4.111	3.889	3.685	3.326	2.951
7	5.389	5.206	5.033	4.868	4.564	4.288	4.039	3.605	3.161
8	6.971	5.747	5.535	5.335	4.968	4.639	4.344	3.837	3.329
9	6.515	6.247	5.995	5.759	5.328	4.946	4.607	4.031	3.463
10	7.024	6.710	6.418	6.145	5.650	5.216	4.833	4.192	3.571
11	7.499	7.139	6.805	6.495	5.938	5.453	5.029	4.327	3.656
12	7.943	7.536	7.161	6.814	6.194	5.660	5.197	4.439	3.725
13	8.358	7.904	7.487	7.103	6.424	5.842	5.342	4.533	3.780
14	8.745	8.244	7.786	7.367	6.628	6.002	5.468	4.611	3.824
15	9.108	8.559	8.060	7.606	6.811	6.142	5.575	4.675	3.859
16	9.447	8.851	8.313	7.824	6.974	6.265	5.668	4.730	3.887
17	9.763	9.122	8.544	8.022	7.120	6.373	5.749	4.775	3.910
18	10.059	9.372	8.756	8.201	7.250	6.467	5.818	4.812	3.928
19	10.336	9.604	8.950	8.365	7.366	6.550	5.877	4.843	3.942
20	10.594	9.818	9.129	8.514	7.469	6.623	5.929	4.870	3.954
25	11.654	10.675	9.823	9.077	7.843	6.873	6.097	4.948	3.985
30	12.409	11.258	10.274	9.427	8.055	7.003	6.177	4.979	3.995
40	13.332	11.925	10.757	9.779	8.244	7.105	6.233	4.997	3.999
50	13.801	12.233	10.962	9.915	8.304	7.133	6.246	4.999	4.000

Appendix 9.4. Annual annuity factors for a present amount $\dfrac{i(1+i)^n}{(1+i)^n-1}$

N	7	8	9	Interest Rate 10	12	15	20
1	1.07000	1.08000	1.09000	1.10000	1.12000	1.15000	1.20000
2	0.55309	0.56077	0.56847	0.57619	0.59170	0.61512	0.65455
3	0.38105	0.38803	0.39505	0.40211	0.41635	0.43798	0.47473
4	0.29523	0.30192	0.30867	0.31547	0.32923	0.35027	0.38629
5	0.24389	0.25046	0.25709	0.26380	0.27741	0.29832	0.33438
6	0.20980	0.21632	0.22292	0.22961	0.24323	0.26424	0.30071
7	0.18555	0.19207	0.19869	0.20541	0.21912	0.24036	0.27742
8	0.16747	0.17401	0.18067	0.18744	0.20130	0.22285	0.26061
9	0.15349	0.16008	0.16680	0.17364	0.19768	0.20957	0.24808
10	0.14328	0.14903	0.15582	0.16275	0.17698	0.19925	0.23852
11	0.13336	0.14008	0.14695	0.15396	0.16842	0.19107	0.23110
12	0.12590	0.13270	0.13965	0.14676	0.16144	0.18448	0.22526
13	0.11965	0.12652	0.13357	0.14078	0.15568	0.17911	0.22062
14	0.11434	0.12130	0.12843	0.13575	0.15087	0.17469	0.21689
15	0.10979	0.11683	0.12406	0.13147	0.14682	0.17102	0.21338
16	0.10586	0.11298	0.12030	0.12872	0.14339	0.16795	0.21144
17	0.10243	0.10963	0.11705	0.12466	0.14046	0.16537	0.20944
18	0.00941	0.10670	0.11421	0.12193	0.13794	0.16319	0.20781
19	0.09675	0.10413	0.11173	0.11955	0.13576	0.16134	0.20646
20	0.09439	0.10185	0.10955	0.11746	0.13388	0.15976	0.20536
21	0.09229	0.09983	0.10762	0.11562	0.13224	0.15842	0.20444
22	0.09041	0.09803	0.10590	0.14401	0.13081	0.15727	0.20369
23	0.08871	0.09642	0.10438	0.11257	0.12956	0.15628	0.20307
24	0.08719	0.09498	0.10302	0.11130	0.12846	0.14453	0.20255
25	0.08581	0.09368	0.10181	0.11017	0.12750	0.15470	0.20212
26	0.08456	0.09251	0.10072	0.10916	0.12665	0.15407	0.20176
27	0.08343	0.09145	0.09973	0.10826	0.12590	0.15353	0.20147
28	0.08239	0.09049	0.09885	0.10745	0.12524	0.15306	0.20122
29	0.08145	0.08962	0.09806	0.10673	0.12466	0.15230	0.20085
30	0.08059	0.08883	0.09734	0.10608	0.12414	0.15230	0.20085
31	0.07980	0.08811	0.09669	0.10550	0.12369	0.15200	9.20070
32	0.07907	0.08745	0.09610	0.10497	0.12328	0.15173	0.20059
33	0.07841	0.08685	0.09556	0.10450	0.12292	0.15150	0.20049
34	0.07780	0.08630	0.09508	0.10407	0.12260	0.15131	0.20041
35	0.07723	0.08580	0.09464	0.10369	0.12232	0.15113	0.20034
40	0.07501	0.08386	0.09296	0.10226	0.12130	0.15056	0.20014
45	0.07350	0.08259	0.09190	0.10139	0.12074	0.15028	0.20005
50	0.07246	0.08174	0.09123	0.10086	0.12042	0.15014	0.20002
55	0.07174	0.08118	0.09079	0.10053	0.12024	0.15007	0.20001
60	0.07123	0.08080	0.09051	0.10033	0.12013	0.15003	0.20000
65	0.07087	0.08054	0.09033	0.10020	0.12008	0.15001	0.20000
70	0.07062	0.08037	0.09022	0.10013	0.12004	0.15000	0.20000
75	0.07044	0.08025	0.09014	0.10008	0.12002		
∞	0.70000	0.80000	0.90000	0.10000	0.12000	0.15000	0.20000

Appendix 9.5. Annual annuity factors for a future amount $\dfrac{i}{(1+i)^n-1}$

N	7	8	9	Interest Rate 10	12	15	20
1	1.00000	1.00000	1.00000	1.00000	1.00000	1.00000	1.00000
2	0.48309	0.48077	0.47847	0.47619	0.47170	0.46512	0.45455
3	0.31105	0.30803	0.30505	0.30211	0.29635	0.28798	0.27473
4	0.22523	0.22192	0.21867	0.21547	0.20923	0.20026	0.18629
5	0.17389	0.17046	0.16709	0.16380	0.15741	0.14832	0.13438
6	0.13980	0.13632	0.13292	0.12961	0.12323	0.11424	0.10071
7	0.11555	0.11207	0.10869	0.10541	0.09912	0.09036	0.07742
8	0.09747	0.09401	0.09067	0.08744	0.08130	0.07285	0.06061
9	0.08349	0.08008	0.07680	0.07364	0.06768	0.05957	0.04808
10	0.07238	0.06903	0.06582	0.06275	0.05698	0.04925	0.03852
11	0.06336	0.06008	0.05695	0.05396	0.04842	0.04107	0.03110
12	0.05590	0.05270	0.04965	0.04676	0.04144	0.03448	0.02526
13	0.04965	0.04652	0.04357	0.04078	0.03568	0.02911	0.02062
14	0.04434	0.04130	0.03843	0.03575	0.03087	0.02469	0.01689
15	0.03979	0.03683	0.03406	0.03147	0.02682	0.02102	0.01388
16	0.03586	0.03298	0.03030	0.02782	0.02339	0.01795	0.01144
17	0.03243	0.02963	0.02705	0.02466	0.02046	0.01537	0.00944
18	0.02941	0.02670	0.02421	0.02193	0.01794	0.01319	0.00781
19	0.02675	0.02413	0.02173	0.01955	0.01576	0.01134	0.00646
20	0.02439	0.02185	0.01955	0.01746	0.01388	0.00976	0.00536
21	0.02229	0.01983	0.01762	0.01562	0.01224	0.00842	0.00444
22	0.02041	0.01803	0.01590	0.01401	0.01081	0.00727	0.00369
23	0.01871	0.01642	0.01438	0.01257	0.00956	0.00628	0.00307
24	0.01719	0.01498	0.01302	0.01130	0.00846	0.00543	0.00255
25	0.01581	0.01368	0.01180	0.01017	0.00750	0.00470	0.00212
26	0.01456	0.01251	0.01072	0.00916	0.00665	0.00407	0.00176
27	0.01343	0.01145	0.00973	0.00826	0.00590	0.00353	0.00147
28	0.01239	0.01049	0.00885	0.00745	0.00524	0.00306	0.00122
29	0.01145	0.00962	0.00806	0.00673	0.00466	0.00265	0.00102
30	0.01059	0.00883	0.00734	0.00608	0.00414	0.00230	0.00085
31	0.00980	0.00811	0.00669	0.00550	0.00369	0.00200	0.00070
32	0.00907	0.00745	0.00610	0.00497	0.00328	0.00173	0.00059
33	0.00841	0.00685	0.00450	0.00292	0.00292	0.00150	0.00049
34	0.00780	0.00630	0.00508	0.00407	0.00260	0.00131	0.00041
35	0.00723	0.00580	0.00464	0.00369	0.00232	0.00113	0.00034
40	0.00501	0.00386	0.00296	0.00226	0.00130	0.00056	0.00014
45	0.00350	0.00259	0.00190	0.00139	0.00074	0.00028	0.00005
50	0.00246	0.00174	0.00123	0.00086	0.00042	0.00014	0.00002
55	0.00174	0.00118	0.00079	0.00053	0.00024	0.00003	0.00000+
60	0.00123	0.00080	0.00051	0.00033	0.00013	0.00002	0.00000+
65	0.00087	0.00054	0.00033	0.00020	0.00008	0.00001	0.00000+
70	0.00062	0.00037	0.00022	0.00013	0.00004	0.00000+	0.00000+
75	0.00044	0.00025	0.00014	0.00008	0.00002		
∞	0.00000	0.00000	0.00000	0.00000	0.00000	0.00000	0.00000

CHAPTER 10

PROJECT ANALYSIS AND SIMULATION MODELS

Many researchers working on livestock systems at the micro level will be called upon to participate in analysis of a livestock project. Regardless of the role played, a basic understanding of "the big picture" in project analysis is useful. Consequently, the major purpose of this chapter is to provide an overview of how a project analysis is carried out and its major parts. Many important details, covered very well in Brown (1979) and Gittenger (1972) among others, are thus left out. A further objective in this chapter is to show how the budgeting tools and formulas presented earlier can be used in project analysis. In fact, although the procedures and idiosyncrasies in determination of the various measures provided by project analysis receive the most attention, the practical reality is that computation of the measures is really quite simple--the hard part is project design and budgeting. In effect, computations in project analysis are a micro problem in a macro context.

Project analysis, originally called benefit/cost (B/C) analysis, is a technique devised to evaluate large projects carried out by public agencies. Project analysis has been used in regional analysis. Project analysis has considerable potential for use in livestock systems research but has received little attention in the farming systems literature. Rather, this technique's major use has been by governments and financial intermediaries such as the Interamerican Development Bank and the World Bank. In the United States, B/C analysis is frequently used in a livestock systems framework for planning land use on large publicly owned land that is designated for multiple uses (U.S. Dept. of the Interior, 1981a, 1981b).

There are two very closely related final analyses carried out when projects are evaluated. One, called the financial analysis, is an assessment of a project's viability from the viewpoint of the individuals or agencies contributing capital and sharing in the financial rewards. Usually, farm-level budgets are prepared for each type of operation affected by the project and totals obtained from which the various project measures, such as the benefit/cost ratio or internal rate of return, are calculated. This is equivalent to the rate of return on all capital invested in the project. That is the approach taken in the example problem presented in this chapter.

An alternative or complementary analysis includes calculating the rate of return on the farm family's own investment after all capital provided by financial sources outside the farm family is accounted for (Brown, 1979). These complementary analyses are also often prepared for agencies that help finance the project or provide services. In each case the objective is to determine the financial impact on the group for whom the analysis is being carried out.

The second major part, called the economic analysis, reflects profitability from the viewpoint of society as a whole. Profitability in this context refers to efficient use of a nation's resources in producing national income. The economic analysis incorporates the information generated in the financial analysis and adds to it a number of social variables. The economic analysis uses measures like the benefit/cost ratio, internal rate of return (IRR), and net present worth, while in the financial analysis the two indicators are cash flow and financial rate of return (just another term for IRR).

Sample Problem, Government Viewpoint

The purpose of this section is to explain how a project is designed and the financial analysis carried out. The approach used is presentation of a proposed ranching scheme in East Africa. Two analyses are carried out. The first is for the project as a whole while the second is from the cattle owners' viewpoint.

Sample Situation--Without Project

This chapter's example is set in an East African country where the government is contemplating clearing brush and trees from 10,000 hectares of alluvial plain to utilize seasonal floodwaters for irrigation of seeded perennial grasses. The cattle owners in the area have traditionally grazed their cattle in a communal fashion even though the livestock are privately owned. As a result, much of this semiarid area is extensively overgrazed, and periodic starvation conditions occur during times of drought.

The range reclamation program is designed for strict cattle inventory regulation, and pasture use would be closely monitored. The program's objective is to provide feed to prevent seasonal weight and starvation losses that take place in nearby areas during periods of low range forage production and also to open up an area now virtually closed to cattle production due to tsetse fly infestation. At present, about 2,000 hectares of woody growth have been cleared, seeded and are being used on an experimental basis in a nearby area. Cost and production data are available from that experiment. The problem is to determine the feasibility, using project analysis techniques, of the expanded program.

Table 10.1. Inventory and cattle production measures from example East Africa range improvement project

Item	Units	Without Project	With Project
Production measures			
Calf crop (weaned)	pct	45	60
Death loss (weaned calves & older)	pct	10	5
Replacement rate	pct	18	18
Land per AU	ha	25	1.5
100 cow basis			
Inventory			
Animal units	AU[a]	152	195
Mature cows	no	100	100
2-yr heifers	no	18	27
Yearling heifers	no	20	28
4-yr steers or bulls	no	15	24
3-yr steers or bulls	no	16	25
2-yr steers or bulls	no	18	27
Yearling steers or bulls	no	20	28
Total	no	207	259
100 cow basis			
Animals marketed or consumed	no		
Cows	no	18	18
Calves	no	--	--
Steers, bulls, & heifers	no	15	33[b]
Total	no	33	51
Total inventory			
Animal units	AU	400	6,667
Mature cows	no	263	3,419
2-yr heifers	no	47	923
Yearling heifers	no	53	957
4-yr steers or bulls	no	39	821
3-yr steers or bulls	no	42	855
2-yr steers or bulls	no	47	923
Yearling steers or bulls	no	53	957
Total	no	544	8,855
Animals marketed or consumed			
Cows	no	47	615
Calves	no	--	--
Steers, bulls & heifers	no	42	1,129
Total		89	1,744

[a]1 mature cow or 2-yr heifer = 0.7 AU, yearling heifer = 0.6 AU, 4-yr bull = 1.0 AU, 3-yr bull = 0.9 AU, 2-yr bull = 0.8 AU yearling bull = 0.7. AU.
[b]24 4-yr steers or bulls plus 9 heifers not needed for

Table 10.2 Financial analysis, before financing, example East African range improvement project, viewpoint of government

	Year				
	1	2	3	4	5-20
	- - - - - - - - - U.S.$ - - - - - - - - -				
Without project					
Gross costs[a]	263	263	263	263	263
Gross benefits[b]	4,323	4,323	4,060	40,060	4,323
Net benefits	4,060	4,060	4,060	4,060	4,060
With project					
Gross costs					
Clearing[c]	300,100				
Fencing[d]					
Seed[e]	120,000				
Seeding[f]					
Dike[g]	6,300				
Water tanks[h]	26,400				
Corrals[i]					
Pasture maintenance[j]	- -	150	300	400	500
Animal maintenance[k]	- -	3,419	3,419	3,419	3,419
Salvage	- -	- -	- -	- -	0
Total	452,700	3,567	3,719	3,819	3,919
Gross benefits[l]	- -	89,937	89,937	89,937	89,937
Net benefits	-452,700	86,358	86,218	86,118	86,018

continued

Table 10.2 Financial analysis, before financing, example East African range improvement project, viewpoint of government (continued)

	Year				
	1	2	3	4	5-20
	- - - - - - - - - U.S.$ - - - - - - - - -				
Incremental					
Gross incremental benefits[m]	4,323	85,614	85,614	85,614	85,614
Gross incremental costs[n]	452,437	3,306	3,456	3,556	3,656
Net incremental benefit (Cash flow)[o]	-456,760	82,308	82,158	82,058	81,958

[a]Gross costs. No grazing costs as the area is used in a purely extractive sense. $1.00 per cow for miscellaneous expenses such as medicines not provided by the government.
[b]Gross benefits.

Animals consumed or marketed	Number	Value or price per kilo - U.S.$ -	Weight per animal - Kilos -	Total weight	Revenue - U.S.$ -
Cows	47	0.15	270	12,690	1,904
Steers or bulls & heifers	42	0.18	320	13,440	2,419
Total				26,130	4,323

[c]Clearing

Type cover	Cost per ha -US$-	Area -ha-	Total Cost -US$-
Heavy	40	3,340	133,600
Medium	25	6,660	166,500
Total		10,000	300,100

continued

Footnotes Table 10.2. (continued)

dFencing. None, cattle are herded all the time.
eSeed. Seed at $1.00 per kilo, seeded at 12 kg per ha.
fSeeding. No cost, done by cattlemen themselves.
gDikes. 63 at $1.00 each. Done with plow and oxen.
hWater tanks. 20 at $2,000 each = $40,000. Estimated that 2/3 of their use will be for open range cattle. Therefore, only 66 percent is charged to the project.
iCorrals. 20 at no cost as done by cattlemen themselves.
jPasture maintenance. Cost primarily for eradication of woody growth using hand methods. No cost first year, 30 percent second year, 60 percent third year, 80 percent fourth year, 100 percent from fifth year.
kAnimal maintenance. $1.00 per mature cow.
lGross benefits.

Animals marketed	Number	Value or price per kilo	Weight per animal	Total weight	Revenue
		- U.S.$-		- - kg - - -	-U.S.$-
Cows	615	0.15	270	166,050	24,907
Steers or bulls & heifers	1,129	0.18	320	361,280	65,030
Total				527,330	89,937

mGross incremental benefits. Gross benefits with project minus gross benefits without project.
nGross incremental costs. Gross costs with project minus gross costs without project.
oNet benefits with project minus net benefits without the project.

The first step is to take high altitude photographs of the project area (free of charge by the military) and sites delineated on them. These sites are later redefined by an on-the-ground inspection, at which time a total of 23 separate sites or pastures are finally delineated. Density of woody growth is used as a measure of forage productivity as well as the basis to estimate clearing cost.

All of the area's users are interviewed to obtain information about current and projected use of the project sites. A social analysis is included to determine cultural factors might preclude the pastures from being used as envisioned by the planners. During the survey it is determined that only 544 cattle of East African Zebu origin are grazed in the 10,000 hectare area due to dense woody growth in much of the area.

The survey indicates that of 544 animals of East African Zebu origin now grazed in the area, 263 are mature cows (Table 10.1). A check on weights indicates that 1 mature cow or 1 two-year old heifer is equivalent to 0.7 AU (in contrast to 1.0 for English breed cattle). Using these coefficients, plus others given in footnotes to the table, it is determined that 400 animal units are now carried on the project's proposed 10,000 hectares, or 25 hectares per AU. It is also determined that the weaned calf crop is 45 percent, death loss is 10 percent, and the cow replacement rate is 18 percent.

Further information gathered in the survey is used to estimate gross costs and returns under the present situation. There are no current expenses estimated for grazing as the area is used in a purely extractive sense, but there is an estimated cost equivalent to $1.00 per mature cow for miscellaneous expenses such as medicines not provided by the government. The gross costs without the project are thus $263 (Table 10.2).

There are currently 47 cows replaced each year, some of which are consumed locally while others are sold. In addition, there are 42 other animals marketed or consumed (Table 10.1). Estimation of the gross benefits, given in the footnotes to Table 10.2, indicates that 26,130 kilos are sold annually. The extraction rate is thus 2.6 kg per hectare. Cattle not sold but consumed locally can be valued at their opportunity value (the revenue they would have fetched had they been sold). With cows valued at $0.15 per kg and other animals at $0.18 per kg, the total revenue, (the gross benefits) are $4,323. The net benefits without the project are thus $4,060 annually.

With Project Cost Estimation

Details on how project costs were calculated are given in the footnotes to Table 10.2. Land clearing is primarily done by hand, with costs for heavier growth calculated at $40 per hectare and medium growth at $25.00. The total cost for clearing is $300,100. There is no fencing cost as cattle are herded at all times and corralled at night. Seed for the project is harvested from a nearby experiment station and is essentially charged at harvesting cost. The proposed cattle owners do the seeding by hand during their spare time so it is assumed their opportunity cost of labor is zero. Thus, there is no charge for this expense item. Dikes (63 at $100 each) are constructed to assist in spreading the floodwater. In addition, 20 dirt tanks are estimated as being needed for stockwater. However, only two-thirds of their cost is charged to the project as cattle grazing on nonproject range are also expected to have access to some of the facilities. Pasture maintenance, primarily for eradication of woody growth, is estimated at $500 annually with a reduced cost in early years of the project. Animal maintenance is still calculated at $1.00 per mature cow. Total cost for the first year, all of which is for investment, is estimated at $452,700.

Estimation of Cattle Inventory and Marketing

The range management specialist on the feasibility study team estimates that beginning in the second year only 1.5 hectares will be required per AU. The animal scientist reckons that through better nutrition the weaned calf crop will increase to 60 percent while the death loss will decline to 5 percent. The problem then, with the cow replacement rate at 18 percent, is to determine total herd inventory and the number of animals marketed.

A relatively easy way to estimate the projected inventory by class of animals is to first place the herd on a 100-cow basis as done in Table 10.1. Because there is a 60 percent calf crop, 30 females are expected. With a 5 percent death loss this leaves 28 yearlings and 27 two-year olds. The same procedure is used to estimate males. A total of 259 animals are thus in the inventory for, and including, every 100 mature cows. This is in contrast to the 207 without the project.

The next step is use the AU projections in combination with the inventory on a 100-cow basis to determine the total number of head by class of cattle. The total of 195 AUs on a 100-cow basis is divided into the 6,667 AUs that can be carried on the project area (10,000 ha ÷ 1.5 AU/ha). This coefficient of 34.19 is then multiplied by each of the classes in the 100-cow inventory and the results summed to determine the total number of cattle (8,885).

The animal scientist determines that 18 percent of the cows should be culled annually so there are 615 cows either marketed or consumed by the livestock owners. There are 821 four-year-old males and 308 two-year-old heifers not needed for replacement that are also available for sale, or a total of 1,129 animals. It is recognized, of course, that in practice sales will fluctuate between years and between classes of cattle, but this format at least provides a means to estimate a stable population.

With Project Benefits

The number of animals marketed is multiplied by the weight per head and price per kg to determine total revenue, which is calculated to be $89,937 annually beginning the second year (footnotes, Table 10.2). The extraction rate is 527,330 kg, or 52.7 kg per hectare, about a twenty-fold increase above the without-project situation. The net benefits for the first year are -$452,700; but they become a positive $86,368 in the second year, although they decrease to $86,018 by the fifth year as an increasing amount of pasture maintenance is required. Net benefits remain at this level for years five-twenty.

Net Incremental Benefits (Cash Flow)
    Gross benefits without the project are subtracted from gross benefits with the project to arrive at the gross incremental benefits. The same procedure is followed for costs, which are then subtracted from benefits to arrive at net incremental benefits, or total cash flow. This is a negative $456,760 in year one, but a positive $82,308 in year two. Total cash flow declines slightly to $81,958 by year five. Cash flow is a good analytical tool to provide an overall view of the project.

Net Present Worth
    A second way to measure the economic viability of a project is net worth, a procedure that involves picking a discount rate that is the best rate that could be earned by the project recipient(s) in investment possibilities outside the project. The discount rate is not necessarily the same as the interest rate at which funds for the project can be borrowed. Let us assume a rate of 12 percent and a project life of 20 years.
    The net incremental benefits from Table 10.2 are carried over to Table 10.3 and multiplied by the appropriate 12 percent discount factors. The result of these calculations is a net present worth estimated to be

Table 10.3.   Calculation of net present worth, financial analysis, example East Africa range improvement project, viewpoint of government

Year	Net incremental benefits	Discount factor 12%[a]	Net percent worth
	- U.S.$ -		- U.S.$ -
1	-456,760	.893	-407,437
2	82,308	.797	65,599
3	82,158	.712	58,496
4	82,058	.636	52,189
5-20	81,958	4.432	363,238
Total net present worth (NPW)			132,085

[a]See Appendix 9.2 and Appendix 9.3. The annuity factor of 4.162 is obtained by substracting the factor for year 4 from year 20 (7.469-3.037).

$132,085. The interpretation of this number is that the present value of the annual stream of net incremental benefits from the project during a 20-year period is $132,085, assuming the money tied up in the project could have earned 12 percent interest.

The use of net present worth as a project measure provides an indication of amount but does not say anything about the rate of return to project investment. The fact that one project has a higher net present worth than another project does not necessarily mean that it uses the funds more efficiently. It could simply be that one project is larger in scope than the other. This problem can be avoided by using the financial rate of return.

Financial Rate of Return

The financial rate of return, the same measure explained in the last chapter but called the internal rate of return at that point, is rapidly becoming the most widely accepted indicator of the relationship of benefits to project outlays. The first step is to multiply the net incremental benefits, (the cash flow) by the 16 percent discount factors (Table 10.4). This rate is arbitrarily chosen as one that appears most likely to yield a net present worth equal to zero. The net present worth is still positive, so a rate of 20 percent is tried next. Now the net present worth is negative, thereby indicating that the financial rate is between 16 and 20 percent.

Interpolation yields a rate of return before financing of 17.2 percent, which means that for every $1.00 tied up in the project, there is a return of about $1.17 or, alternatively, 17 percent, not including cost of debt servicing. The interpolation process to determine the rate of return should not span more than 5 percentage points because the interpolation is a linear function while the actual rate is a curvilinear function.

Financial Capability

The calculations indicate that as specified the project is viable, at least from the government's point of view. But no mention has been made of who is going to make the initial investment, how it will be repaid, and the impact of debt servicing. Any project analysis will have a section on these aspects along with a cash-flow diagram showing loan repayments. For example, an agreement may be worked out with a lending agency for the government to fund 20 percent of the investment and for an international agency to fund the remaining 80 percent. The question then is the extent to which borrowers can repay the loan. This is determined by the cash flow analysis, debt servicing, and, after financing, financial rate of return.

Two plans are now examined, one based on the initial situation presented and one that assumes cattle must be purchased to stock the improved area. Both analyses are carried out from the government's point of view.

Table 10.4. Calculation of financial rate of return, example East Africa range improvement project, viewpoint of government

Year	Net incremental benefits (cash flow)	16 percent		20 percent	
		Discount factor[a]	Net present worth	Discount factor[a]	Net present worth
	- - U.S.$ - -		-U.S.$-		-U.S.$-
1	-456,760	.862	-393,727	.833	-380,481
2	82,308	.743	61,155	.694	57,122
3	82,158	.641	52,663	.579	47,569
4	82,058	.552	45,296	.482	39,552
5-20	81,958	3.131	256,610	2.281	186,946
Net present worth			21,997		-49,292

[a]Years 1-4 from Appendix 9.2. Years 5-20 from Appendix 9.3. Factor for years 5-20 is factor in year 20 minus factor in year 4.

$$\text{Financial rate of return} = 16 + 4 \frac{21,997}{(21,997 + 49,292)} = 17.23$$

Plan I

This plan is based on the previous financial analysis assumption that cattle are brought to the project area by producers from their own herds and thus are not an investment cost. Arrangements are worked out with an international lending agency to loan the government 80 percent of the investment cost with a 1-year grace period (this assumes that year one of the project covers a full year) and a 12 percent interest rate. An annual repayment of $49,254 is the product of taking 80 percent of the investment cost ($452,700) to arrive at the amount borrowed (Table 10.5). That figure ($362,160) is then multiplied by 0.136, which is the equal payment amortization factor for 19 years at 12 percent interest (Appendix 9.4). The result, $49,254, is added to the basic costs developed earlier to arrive at the with-project gross benefits.

Net incremental benefits are calculated in the way described in the before financing alternative and transferred to Table 10.6, where the financial rate of return is calculated. Results show that a rate of 34.6 percent is expected under these assumptions. Assuming that the social rate of return (as measured by opportunity cost of capital to government) were about 12 percent, then the project is viable and should be conducted unless there are other projects with a higher rate and there is insufficient capital to fund all of them.

Plan II

Assume that cattle are not available to stock the project area and that they must be purchased at a cost of $308,894 (Table 10.5). Preliminary analysis indicates that the project is not viable under the preceding financing conditions, so "soft-loan" type financing conditions of 6 percent interest and 2 years of grace period are specified. If such an agreement were successfully negotiated, the financial rate of return would be 21.4 percent (Table 10.6). This rate is considerably less than the 34.6 percent when cattle are not included, but 21.4 percent is still above the social cost of capital.

Inflation

One way to cope with inflation is to inflate all costs and returns by some projected rate of inflation. But this cumbersome procedure can be avoided by assuming that prices on both the benefit and cost side will rise at the same rate and thus that their relative values will not change. This method is possible because the livestock analyst's objective is to determine a rate of return, not debt servicing.

Loan repayment is a serious problem for borrowers and lenders because money will lose its value over time in an inflationary economy. If money is lent at a fixed rate, borrowers have a windfall gain with inflation. Some countries will arbitrarily increase the rate of interest to compensate, while others will adjust repayments according to some measure of inflation. The important point for the livestock analyst is to

Table 10.5. Alternative plans, financial analysis, after
financing, example East Africa range
improvement project

	Year				
	1	2	3	4	5-20

- - - - - - - U.S.$ - - - -
Financing, Plan I

	1	2	3	4	5-20
Without project					
Net benefits[a]	4,060	4,060	4,060	4,060	4,060
With project					
Gross costs[a]					
Basic[b]	452,700	3,569	3,719	3,819	3,919
Debt servicing[c]	-	49,254	49,254	49,254	49,254
Total	452,700	52,823	52,973	53,073	53,173
Gross benefits					
Basic	-	89,937	89,937	89,937	89,937
Loan receipts	362,160	-	-	-	-
Total	362,160	89,937	89,937	89,937	89,937
Net benefits[d]	-90,540	37,114	36,964	36,864	36,764
Net incremental benefits[e]	-90,600	33,054	32,904	32,804	32,704

Financing, Plan II

	1	2	3	4	5-20
Without project					
Net benefits	4,060	4,060	4,060	4,060	4,060
With project					
Gross Costs					
Basic	452,000	3,569	3,719	3,819	3,919
Cattle[f]	308,984	--	--	--	--
Debt servicing[g]	--	--	56,002	56,002	56,002
Total	770,894	3,054	59,721	59,821	59,921
Gross benefits					
Basic	--	89,937	89,937	89,937	89,937
Loan receipts	608,715	-	-	--	--
Total	608,715	89,937	89,937	89,937	89,937
Net benefits	-162,179	86,368	30,216	30,116	30,016
Net incremental benefits	-166,239	82,308	26,156	26,056	25,956

understand how inflation is dealt with by lending agencies so that
appropriate advice can be given on project design.

Producer's Viewpoint

The analysis carried out thus far has been from the government's
viewpoint. But what about the producers themselves. Will they
participate? Is the project in their interest? As pointed out earlier,
noneconomic factors would have to be carefully analyzed. Assuming that
analysis indicates producers would be interested and would participate
along the physical lines outlined, the question then rests on the economic
benefits to producers. Following are a number of scenarios.

Scenario 1. Cattle Brought from Other Areas and Given to Cattle Owners
No further economic analysis is required in this case because the
cattle owners are given a free resource. It is a gift whose cost is paid
by the government.

Scenario 2. Cattle Brought from Other Areas and Cattle Owners Pay 80 Percent of Investment
This scenario assumes that producers are required to make a 20
percent downpayment and obtain a loan for 80 percent of the investment.

Footnotes Table 10.5
[a]From Table 10.2.
[b]From Table 10.2, with project gross costs.
[c]$452,700 times 0.80 - $362,160 loan receipts, times 0.136
(factor to determine annual cost for 19 years at 12
percent, Appendix 9.4).
[d]With project gross benefits minus gross costs.
[e]With project net benefits minus without project net
benefits.

[f]Class of cattle	Number	Cost per unit	Total cost
Mature cows	3,419	41	140,179
2 yr heifers	923	40	36,920
Yearling heifers	957	25	23,925
3 yr steers or bulls	855	55	47,025
2 yr steers or bulls	923	40	36,920
Yearling steers or bulls	957	25	23,925
Total			308,894

[g]$760,894 (452,000+$308,984) times 0.80=$608,715 loan
receipts times 0.092 (factor to annualize cost 18 years at
6 percent interest, Appendix 9.4).

Table 10.6. Calculation of financial rate of return,
            example East Africa range improvement project,
            viewpoint of government

PLAN I

Year	30 percent Net incremental benefits	30 percent Discount factor[a]	30 percent Net present worth	35 percent Discount factor[a]	35 percent Net present worth
	-U.S.$-		-U.S.$-		-U.S.$-
1	-94,600	.769	-72,747	.741	-70,099
2	33,054	.592	19,568	.549	18,147
3	32,904	.455	14,971	.406	13,359
4	32,804	.350	11,481	.301	9,874
5-20	32,704	1.150	37,610	.853	27,897
NPW			10,883		-822

$$\text{Financial rate of return} = 30 + 5 \ \left(\frac{10,833}{10,883 + 822}\right) = 34.6$$

PLAN II

Year	20 percent Net incremental benefits	20 percent Discount factor[a]	20 percent Net present worth	25 percent Discount factor[a]	25 percent Net present worth
	-U.S.$-		-U.S.$-		-U.S.$-
1	-166,239	.833	-138,477	.800	-132.991
2	82,308	.694	57,122	.640	52,677
3	26,156	.579	15,144	.512	13,392
4	26,056	.482	12,559	.410	10,683
5-20	25,956	2.281	59,206	1.592	41,322
NPW			5,554		-14,917

$$\text{Financial rate of return} = 20 + 5 \ \left(\frac{5,554}{5,554 + 14,917}\right) = 21.4$$

[a]Discount factors not available in appendix tables. See
some other source of discount factors.

The analysis would be carried out in the same way as the government viewpoint scenario presented earlier except that an additional section would be required about source of the 20 percent equity.

Other Scenarios

A number of other scenarios could be developed around various financing schemes, but the preceding scenarios are sufficiently diverse to explain the method and approach required. In these examples government might obtain partial financing from an international agency and then reloan the funds to the producers, probably at a higher rate to cover administrative costs and repayment risks.

Economic Analysis

The examples given earlier are known as financial analyses because they are presented from the producer's or project originator's viewpoint. But the likely effects of proposed projects on the primary clientele and other specific groups are only part of a project's potential impact. Development agencies are also concerned that projects make a positive contribution to national economic growth and, as such, a so-called economic analysis is also usually carried out. This analysis is based on the same basic kinds of information used in the financial analysis, but some of the items included as benefits and costs and the values placed on them may be different. The purpose of this section is to clarify some of the differences between the analyses.

Costs and benefits in the economic analysis, just as in the financial analysis, are discounted. But if a net present worth or a benefit/cost ratio is sought, the discount rate will be the social rate rather than the opportunity rate. Basically, the social rate of return is the rate that capital resources in the nation as a whole could produce if not used for this particular project.

The economic analysis will also contain the aggregate or complete farm-level effects. Furthermore, it includes indirect or secondary effects. For example, from a nationwide viewpoint, taxes that producers would pay on the new project are transfers, not costs, as they just move from one group in the economy to another group. Participation subsidies and interest charges on domestically borrowed funds used in the project are also transfers and not drains on the economy. But any technical help, such as extension agents, are additional costs if this means less attention to other problems.

Calculation of the benefit/cost ratios is presented in Table 10.7, assuming no change in gross incremental benefits and costs from those presented in Table 10.2 because the objective is simply to show the method for calculating the ratio rather than differences in items included in the economic versus the financial analysis. Each of the costs and benefits in Table 10.7 are multiplied by the discount factor chosen to represent social costs, 12 percent in this case. Division of the total

Table 10.7. Calculation cost ratio, East African range
improvement project, economic analysis, before
financing

Year	Gross incremental benefits[a]	Gross incremental costs[a]	Discount factor 12%[b]	Present worth Benefits	Costs
	- U.S.$. -	- U.S.$. -		- - U.S.$ - -	
1	- 4,323	452,437	.893	-3,860	404,026
2	85,614	3,306	.797	68,234	2,635
3	85,614	3,306	.712	60,957	2,461
4	85,614	3,556	.636	54,451	2,262
5-20	85,614	3,656	4.432	379,441	16,203
NPW				559,223	427,587

Benefit/cost ratio $= \dfrac{\$559,223}{\$427,587} = 1.31$

[a]Table 10.2.
[b]Appendix 9.2 and 9.3.

present worth of gross incremental benefits by total and present worth of gross incremental costs results in the benefit cost ratio, 1.31 to 1 in this case. The interpretation is that for every $1.00 tied up in the project (periodic annual costs are also included), there is a return of $1.31.

The benefit/cost analysis decision rule is to accept all projects with a ratio greater than 1.0 to 1 and to accept the project with the highest ratio. But, as we explained earlier, benefit/cost ratios can lead to incorrect ranking of projects and, in addition, require that a discount rate be chosen. That choice can lead to political difficulties (because the rate chosen has a major impact on a project's viability), and consequently, the financial rate of return or the economic rate of return has become the preferred indicator. The economic rate of return (not shown) is calculated in the same manner as the financial rate; only the name changes. It would be calculated on the revised cash flows in the economic analysis.

The economic analysis aims to determine the real worth to the national economy of project inputs and costs. Because prices are often distorted by imperfect competition, price ceilings, resource immobility, and so forth, it is often necessary to make adjustments in prices of labor, land, world-traded commodities, and exchange rates. Estimation of the real prices, called shadow prices, is much debated by economists. But the main point, as in all aspects of project analysis, is to use values that correspond to the project's focus and to be consistent from project to project. The same thing applies to estimation of secondary benefits that are spinoffs or indirect activity generated as a result of the project.

Simulation as a Livestock-Analysis Technique

Development of livestock projects can be a laborious procedure, especially because physical data are seldom available. One approach to analyze potential projects is by simulating them with a model. Simulation of systems also has considerable relevance for a wide variety of other livestock research work. Consequently, a basic discussion about the technique and its application at the micro level are now provided. Micro-level models about livestock are discussed elsewhere (M'Pia and Simpson, 1984; Sullivan, Farris and Simpson, 1984; White et al., 1978).

Mathematical and computer modeling have been applied with ever greater efficiency and sophistication to livestock production systems since the 1960s, when large main frame computers first became readily available (Levine and Hohenboken, 1982). Simulation is the method by which experimental information about systems, or models of systems, is generated (Johnson and Rausser, 1977). Simulation is widely used in the formulation, evaluation and application of livestock systems models. The procedures range from those carried out by hand calculations to intricately structured models designed for wide usage that require large computers.

Sensitivity analysis, a process of gathering information with which to evaluate alternative courses of action, is one simple variant of simulation analysis frequently used by livestock systems researchers. In simulation, sensitivity analysis is used to test the robustness of decisions taken in constructing and applying models. As a consequence, interest is directed to areas where doubt, uncertainty and ignorance are greatest. Models very sensitive to changes in assumptions about which there is substantial uncertainty warrant skepticism as to applicability. Sensitivity analysis is especially useful in project analysis.

One area that has received considerable attention in livestock systems simulation is response surfaces--that is, the functional relationships between parameters, environmental variables, structured choices and, to some extent, criterion functions and internally determined variables. The literature on response surfaces contains well-developed procedures that have potential for systems studies, but invariably there is a dearth of data that can be directly applied (Heady and Bhide, 1983).

The purpose of simulation models, by their very nature, is generally not in forecasting or predicting, although some models of this type have been constructed. Rather, the purpose is projecting, with an interpretation being, for instance, "if x were to change, y would be the result." In effect, simulation is fundamentally a tool to describe a real world situation through a model. Usually, the model will then be used to evaluate changes in the variables on parameters.

Some of the more important work on micro-level livestock and pasture simulation modeling was spearheaded by Dr. Tom Cartwright at Texas A&M University beginning in the early 1970s. As a result, the Texas A&M Cattle Production Systems Model, a deterministic model to simulate beef cattle production under a wide range of management schemes and environments with cattle differing widely in genotype for size, growth and milk production, has become the basis for most simulation models. The current A&M model is a unique approach because it simulates levels of performance from specified feed resources and cattle production potentials. In contrast, earlier herd production models utilized cattle performance as input data and then simulated the requirements for these performance levels (Sanders and Cartwright, 1979a, 1979b).

Review of developments that have taken place in simulation models not only provides an excellent source of information for researchers, but is also an important segment in the history of ideas and their formulation. Some of the early work was use and validation of the fertility section by Sanders (1974). He then (1977) used the model to simulate production of cattle differing widely in genotype for size and milk production under two different sets of central Texas conditions. About the same time Davis, Cartwright and Sanders (1976) used the model to evaluate alternative management strategies in two different regions of Guyana. Nelson, Cartwright and Sanders (1978) then simulated production efficiency for cattle differing in genetic potential for size and milk production under three management systems in Central Texas. Ordonez (1978) applied the Texas A&M model to the western high plains of Venezuela.

Cartwright, Gomez, Sanders and Nelson (1977) modified the original model to simulate production systems in Colombia where cows are milked. The same modification of the model was used to simulate dual-purpose cattle production in Botswana (ILCA, 1978). Notter (1977) modified the model to simulate crossbreeding systems and to compare simulated performance of cattle with different potentials in different breeding systems.

Another simulation model was developed to study voluntary forage intake, energy requirements for maintenance, liveweight change and calving rate for grade Zebu cows in the Llanos of Colombia (Levine, Hohenboken and Nelson, 1981; Levine and Hohenboken, 1981). These researchers constructed a frequency distribution relating liveweight at time of mating to subsequent calving rate, a procedure that was then used to simulate the calving rates. The model is used to predict the amount of improved pasture or supplemental feed necessary to raise calving rate, a procedure that was then used to simulate the calving rates. The model is used to predict the amount of improved pasture or supplemental feed necessary to raise calving rates to various levels. This model was fitted by computer and then adapted to programmable calculators.

One response-surface-type simulation model designed to use a whole farm simulator in concert with animal enterprise models is CANPAS, a pasture production model designed to study irrigation management in New Zealand (Fick, 1980). Other work on pasture simulation models has been done by Fick (1977), Hunt (1977), Keulen (1975), and Peake, Henzell and

Stirk (1979). McCown (1980) developed a simulation model to evaluate climatic-related aspects of pastoral lands in tropical Australia. This work included in-depth research on characterizing these seasons, using agroclimatic variables derived from a weekly pasture growth index, as to their quantitative relation to cattle liveweight changes (McCown, Gillard, Winks and Williams, 1981a, 1981b).

One of the more complicated biological models is the one developed at the ILCA by Panos A.S. Konandreas and Frank M. Anderson (1980) for use in describing African livestock systems. This model is based on monthly time steps in which the status of each animal is determined by a set of biological processes and decision rules embedded in the model. Many other cattle-production-oriented models have been developed or based on the Texas A&M model (Abassa, 1984; Boyd and Koger, 1974; Bravo, 1972; Ive, 1976; and Long, Cartwright and Fitzhugh, 1975).

Another model about range use, entitled RANGES 1, mimics the biological behavior of livestock grazing on grassland with the objective of generating management information (Innis and Miskimins, 1973). A dynamic model to simulate beef production with various breeds and crossbreds was constructed by Congleton and Goodwill (1980a) in the simulation language DYNAMO. The value of simulation was shown by their application (1980b and 1980c) of the model to cow-culling strategies. Herd growth models are reviewed by Hallan (1983). Simulation of physical parameters is a very useful tool for a wide variety of problems. Simulation is also a prerequisite to understand how a system operates and what the physical consequences of changes are. But policymakers, and producers in particular, are primarily concerned about financial aspects of changes in physical parameters. As a consequence, a number of financially oriented models have been developed.

One study, done at the ranch level based on the Texas A&M biological model, was designed to answer the questions, what type of beef animal is most profitable, and can profits be increased by maintaining ownership of calves through the stocker and feeding stages (Stokes, Farris and Cartwright, 1981)? The study compares, in a long-run setting, costs and returns associated with 9 beef herds differing in mature size and milking potential. Partial budgets were used to evaluate the profitability of each animal class.

Another approach to evaluate rancher management decision-making was taken by Angirasa (1979) by interfacing a simulation and two linear programming models. He first used the TAMU (Smith, 1979) biological model to simulate beef cattle production for four livestock enterprises (cow/calf stocker, drylot/finishing, and forage finishing) on three alternative forage systems and with four levels of winter feeding. Then, using the simulated results as model inputs, a linear programming model was used to determine the least-cost acreage of a particular forage and purchased hay and supplemental feed required to support a simulated herd. The results generated by these two models were then used as data in an economic model. A program to evaluate alternative long-term ranching level strategies is BEEFLOAN, a herd-growth-based computer financial simulation program with a 10-year planning horizon (Alderman, Simpson and Carlton, 1984).

Another simulation model, primarily targeted to technological evaluation and its relation to prices and credit condition, is a model developed at the Centro Internacional de Agricultura Tropical (CIAT) in Cali, Colombia (Juri, Guiterrez and Valdes, 1977). This model is important because it incorporates diverse risk elements by probabilistic treatment of some parameters, such as calving, death rate and pasture establishment. The model also contains a separate treatment of cattle price cycles through the use of spectral analysis. The model is designed for use in extensive production systems that do not store forage and in which the only activity is cattle raising. The projection period is 25 years. Output includes cash flow and rate of return in addition to physical relationships. Only limited attention is given to expense items. As with all project analyses, data are in present prices.

The CIAT model was used by Sere and Dopler (1980-81) to analyze the economics of large-scale beef production alternatives in Togo. Sere and Dopler varied breed and pasture type for different levels in efficiency in this part of the western African savanna belt. It is instructive that the most profitable systems were extensive ones (similar to nomadic systems) with minimum investments in fencing and improved pastures. A very good recent summary of many models is the book by Spreen and Laughlin (1986).

Numerous decisionmaking programs for microcomputers have been developed at U.S. land grant universities over the past several years. There are also many available through private companies. In general these all deal with record keeping or analyses, and economics of simulated consequences. The late 1980s is truly the era of computers in developed countries. They are becoming common, and in many cases indispensible, for livestock related work in LDCs.

CHAPTER 11

FARMING SYSTEMS RESEARCH AND
EXTENSION AND SWINE SYSTEMS

Considerable empirical evidence has indicated that the needs of small
farmers (essentially those near the subsistence level and whose operations
are only marginally connected with the market economy) are not
adequately met (Khan, 1978; Poleman and Freebairn, 1973). As a consequence, the U.S. Congress in the late 1960s mandated that the U.S.
Agency for International Development specifically address the needs of
small farmers rather than just waiting for economic benefits to "trickle
down" to them (Flora, 1982; Shaner, 1981). Many other donor countries
followed a similar development philosophy (Harwood, 1979). This redi-
rected emphasis brought researchers and development specialists to the
realization that very little was known about the decision-making processes
of small farmers and the way in which they manage their inputs and
resources (Norman, 1980). The new emphases led to development of
farming systems research and extension (FSR/E) methodology.

One purpose of this chapter is to tie major concepts set forth in
FSR/E with some of the specialized economic tools and concepts available
to researchers working on livestock problems. A second objective is to
explain how those concepts as well as budgeting can be used to analyze
swine systems in developing countries.

The subdiscipline of farm economics was quite well developed by the
mid- to late 1960s. With the advent of computers, linear programming
seemed to be the ultimate tool to analyze farming problems and to
optimize resource combinations. But as development specialists delved
into problems of small farmers, especially at the subsistence level, it
became apparent that farm management techniques as they were being
taught lacked something for most types of agriculture in less developed
countries because of the failure to account for the intimate linkage
between the unit of production--the farming system (FS)--and the
consumption unit (farming household) (Byerlee, Harrington and
Winklemann, 1982; Norman and Gilbert, 1981). In effect, it appeared that
insufficient attention was being given to the way members allocate

managerial know-how and resources to crops, livestock raising, the home and off-farm enterprises in order to attain goals.

The term farming systems was used during the 1970s to identify activities with a common thread or purpose. Then, by the 1980s, the generic term farming systems research (FSR) came into common use. It then became evident that the two basic components, when used together, constitute the approach. This concept is similar to the one used by Shaner, Philip and Schmehl (1981), who termed it FSR&D (farming systems research and development). The two complementary components of FSR&D, recognized by Norman (1982) using slightly different terminology are the farming systems approach to intrastructural support and policy (FSIP) and the farming systems research and extension approach to technology generation, evaluation and delivery.

FSR is different from traditional farm management research in that (1) it is oriented toward the needs of small farmers; (2) it specifically sets research priorities that reflect the holistic perspective of the whole farm/rural household and natural and human environments; and (3) it is aimed at defining general farming systems in a region under study as a means to identify and describe a population's activities. In addition to farm management, FSR has antecedents in the community development and subsistence-level agricultural programs of the post-World War II period (Holdcraft, 1978). FSR is also more of a social-institution-oriented approach than farm management because greater recognition is given to changes in nontechnical factors, such as markets, pricing policy and infrastructure (Norman 1982). Another characteristic is an interdisciplinary orientation with farming systems teams often being comprised of anthropologists, agronomists, animal scientists and others as well as agricultural economists.

Extension has become recognized as an integral part of the research process because (1) FS research is carried out on farms in collaboration with experiment stations; (2) it is specifically action (change) oriented; and (3) the research approach is to conduct on-farm trials and to measure both crop response and farmers adoption of interventions (Figure 11.1). As a consequence, given that extension personnel also comprise an integral part of the interdisciplinary team (Hildebrand, 1982), the approach has become known as FSR/E. Because the extension part is so important, both from a research as well as action viewpoint, the approach will be called FSR/E for the rest of this book.

The farming systems methodology was specifically designed for use in small-farmer situations. But it is likely that the on-farm testing method may become one of the methodology's major contributions. This is one aspect highlighted throughout this chapter. In this line of thinking, the household unit thus becomes one component to be reckoned with rather than the central focus it has sustained in so much of the FSR/E literature (Norman and Gibbs, 1979; Norman, 1982). Nevertheless, a crucial point is to recognize that producers are the focal point of any livestock operation and that any intervention, to be successful, must take their

225

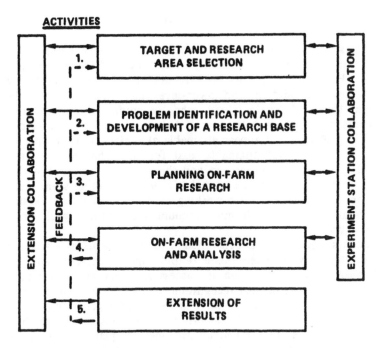

SOURCE: ADAPTED FROM SHANER, PHILIPP AND SCHMEHL, 1981.

FIGURE 11.1   THE FIVE BASIC ACTIVITIES IN FSR/E.

goals into account. This is a key factor in developing research methodo-logy for livestock on-farm trials (Nordblom, Ahmed and Potts, 1985).

Producer Goals

Producers have plans that are developed to achieve certain goals. These goals are often multiple, possibly conflicting, and frequently poorly articulated. For example, farmers and livestock owners have higher-level goals (educate their children) and lower-level, more specific goals (plant at the right time). Gladwin (1980), who has studied this topic exten-sively, points out that lower-level goals help achieve higher-level ones. If a specific goal does not appear attainable at a given time, people will substitute a different yet attainable goal for it. The substitution of lower-level goals in the shorter run and the reinterpretation of higher-level goals in the longer run make real-life decisionmaking seem messy, especially to researchers who are working at the farm level on livestock systems research and as such must account for multiple goals and a multitude of decisions.

Livestock systems analysts concerned with investments as a means to improve productivity, efficiency and output must understand and take producer goals into consideration if successful interventions are to be designed. The problem of goal recognition, and decision about tradeoffs between goals, is complicated, but there are a few goal-generating devices people use that help simplify the identification process. For one thing, most routine decisions are handled by subconscious plans or scripts that are determined by goals (role playing). Also, role playing is an important aspect in development of goals (Schank and Abelson, 1977).

Recognizing that role playing does exist facilitates the researcher's job of identifying constraints and opportunities for change and also helps identify who will and can do what. Another advantage of such recogni-tion is to understand that roles help reduce "cognitive overload" by automatically producing goals, which in turn determines plans or scripts for people. In effect, self-identification of a role cuts down on the number of conscious or "attentive" decisions that must be made in order to survive.

The concept of role playing is important to determine optimal constraints on input mixture in animal agriculture for output depends greatly on availability of individual farm resources and the interest of a particular farm family. For example, in an LDC a farmer with a young family may find goats a lucrative business if labor is available to herd the goats and for ancillary activities such as milking, distributing the milk, tanning hides, making goatskin products, and merchandizing the products. But an older person lacking these resources will probably find the cost of hiring labor excessively expensive compared with the income and thus may feel that raising replacement dairy cattle heifers is a better alternative, even though gross income is reduced.

An important application of goal recognition in small-farm livestock systems work is simply to recognize that decisions involve them, or her or her and him, not just him. A second point is that the farm family as a group must be brought into any project design to understand why certain practices are carried out. Quantitative information is useful, but it is not a substitute for good understanding of what the farmer is thinking. In effect, even though family goals are important on larger operations, they increasingly become a focal point as research is more closely tied to the subsistence level. Much of the information about goals can be obtained in a rapid rural appraisal (Beebe, 1985).

Importance of Animals on Small Farms

Animals form an integral and fundamental part of small-farm systems in LDCs (DeBoer and Welsch, 1977; Groenwold and Crossing, 1975). Except for a few countries, 85 percent or more of the ruminants (buffalo, cattle, sheep and goats), and an even higher proportion of donkeys and horses, are on small farms (McDowell and Hildebrand, 1980, p. 63). Animals are an important part of cultural patterns; at the same time, animals provide economic returns from manure, traction, transport, investment, insurance, fuel, byproducts, skins and hides as well as milk, meat and progeny (Berleant and Schiller, 1977; Duckman and Mansfield, 1971; Sprague, 1976).

Small farmers carry on their integrated crop/livestock operations in ways that are efficient within the constraints set by the traditional societies and environments within which they live (Harris, 1974). But due to rapid population growth, continually higher proportions of people living in urban areas, and an escalation of desires for higher per capita incomes, it is incumbent upon development planners, researchers and others to develop and adapt new and relevant technology to facilitate adjustment. The appropriateness of technology depends on goals and priorities set by producers.

According to McDowell (1979), the usual order of priorities for keeping animals on small, integrated crop/livestock farms is to (1) reduce risks from cropping, (2) accumulate capital, (3) render services, for example, traction, fertilizer and fuel, (4) satisfy cultural needs, (5) ensure status or prestige, (6) provide food, and (7) generate direct income. The key here is that although the ranking will change in different situations, the lowest priorities of small direct producers for keeping animals are provision of food and generation of income. In general, attention is not given to obtaining a high rate of output in animal products because other goods and services are more important; nonanimal factors command more time and effort both in a physical and decisionmaking mode on integrated farms. Futhermore, there must be recognition of the inherent or subconscious effort to keep as many animals as possible because numbers count when animals are viewed as a type of insurance, a source of prestige, and the origin of services.

Conceptualization of Small Farm Systems
Using a Linear Programming Framework

One way in which the small livestock producer situation can be conceptualized is by setting it in a linear programming framework of the type developed in Chapter 3. The problem is graphically portrayed in Panel A of Figure 10.2 in which a small farmer is assumed to have 5 hectares of land that can be used either for crops (vertical axis) or livestock (horizontal axis). A straight line intersects the points to indicate the 5 hectares on each axis that denote the land constraint. A credit constraint absolutely prevents use of more than 2.5 hectares for animal agriculture or 6 (if it were available) for crops. Training and managerial ability of the farm family are identified as labor constraints that only permit 3.0 hectares to be used in livestock or 7 in crops. The result is that this producer is bound by constraints that permit production within the boundaries of OABC. Each point constitutes a potential corner solution.

Suppose that a team visits the area and decides credit is the major constraining factor to improved animal production. More credit is subsequently provided, which permits the producer to potentially use 3.5 hectares for livestock. All 5 hectares could still be cropped as there is no increase in crop-oriented credit. The credit constraint is now "flatter" (Panel B). The team realizes that management ability for animals is higher than the credit constraint is, so no training component is added and that constraint line thus remains unchanged. The corner solutions are now the boundaries of OADE (Panel B).

Now that all the constraints have been plotted, the profit-maximizing combination could be recalculated. The most profitable corner, although not computed for this example, would be determined by calculating income and cost at each corner. If cropping is quite profitable, it may be that corner A will continue to be the optimum solution in both cases. But if this is not the case, the new corner solution will either be D or E, implying that livestock use will be either 1.5 or 3 hectares. Without credit it would have been either 0.7 or 2.5 hectares depending on the situation. It is important to note that the calculations can be carried out ex-ante to determine feasibility of a credit program.

Panel B of Figure 11.2 indicates that if livestock is very profitable relative to crops, the new corner solution would be E, in which case management is the constraining factor. No land would be used for crops as management would not be able to handle both activities. An example might be a family raising sheep in which the children tend them, the wife weaves the wool into sweaters, and the husband sells the sweaters on a street corner in a nearby city.

Now suppose that the team determines the farm family in question has the ability to become better managers and a training component is added. Conceptually, the result would be a "flatter" management constraint line, perhaps extending out past the land constraint (not shown).

Credit (Panel B) would now become the corner solution for all livestock. If sufficient credit and training were both added, the corner solutions would now become either A or G. Provision of any training or credit that would carry the family beyond that point would be wasted as only 5 hectares are available (this only applies to the short run and assumes, of course, that off-farm alternatives are not available).

Another practice that can be graphically (or mathematically, for that matter) shown with linear programming is keeping livestock for prestige, status, insurance, or saving purposes. These conditions form an integral part of the role-playing mechanism that was described earlier. In economically developed societies the need for prestige frequently manifests itself in ownership of expensive automobiles or large homes. In subsistence level livestock economies the need for prestige might be met in some cultures by having multiple wives or owning the greatest number of cattle possible (Little, 1984; Schneider, 1984). Sometimes, of course, livestock are owned simply because people like to have them around, in other cases, it is an effort to maintain a certain image.

Probably the main reason small, integrated crop/livestock producers keep cattle or buffalo is, as pointed out a few pages earlier, to reduce risks from cropping. In effect, livestock are a form of insurance or savings. They can be sold rather easily to generate income, yet are better than keeping cash on hand as sale of them does require some planning and effort, thus acting as an effective deterrent to spur-of-the-moment purchasing. Although there are many noneconomic, or at least only marginally economic, aspects to livestock ownership, the term prestige/savings factor will be used to describe these noneconomic.

The desire by landowners or operators, and smallholders in particular, to keep some livestock regardless of the cost and returns situation is presented conceptually in Figure 11.3, which has been superimposed on the traditional system drawn as Panel A in the previous figure. But in addition, there is a prestige/savings factor that slopes upward to the right because the number of livestock may increase without using additional land. In effect, for many smallholders, the prestige/savings constraint might be hypothesized to have a positive slope indicating more cropping as the amount of land available increases.

A vertical line called "livestock maximum" has also been drawn in at 2 hectares of livestock to indicate this is the maximum possible use of land for livestock regardless of benefits, costs, prestige/savings factor, or other decisionmaking criteria because cropping is the family's source of livelihood. The corner solutions are now O, H, I, J, and K.

Corner H is infeasible as selection of it would mean only 3 or 5 available hectares of land would be used, all of it would be in crops, none of the constraints would be "bumped," and the condition of at least 2 hectares devoted to livestock would not be met. Corner K is also not feasible as this would mean there is no cropping. The optimum solution is thus either corner I or J. At corner I, only 3.5 hectares are used because credit has become such a constraint that even though the

231

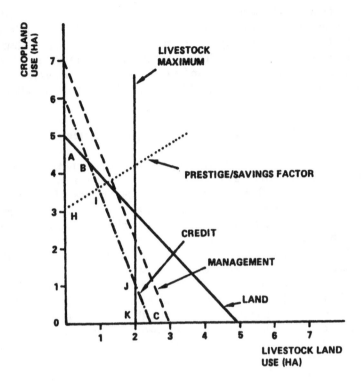

FIGURE 11.3   ADDITION OF PRESTIGE/SAVINGS FACTOR IN GRAPH-
ICAL REPRESENTATION OF LINEAR PROGRAMMING
FRAMEWORK IN AN INTEGRATED SMALL FARM
CROP/LIVESTOCK SYSTEM.

producer wants more livestock, and plans for them, they cannot be obtained or raised. Consequently, only 1.25 hectares of the desired 2 hectares are actually utilized. If, on the other hand, the optimum solution is corner J, then only 1 hectare of land is cropped, but 2 hectares are used for livestock. This situation might prevail in a very dry area where the only way some land can be utilized for livestock is by irrigation. Thus, with restricted credit, part of the land is just allowed to lie fallow.

Small-Farmer Integrated Crop/Livestock Systems

An agricultural system consists of a restricted number of major (dominant) enterprises, crops and types of livestock, and numerous minor ones. As an example, Table 11.1 contains descriptions of integrated small-farm livestock/crop systems found in Asia. A scanning of this table provides a panorama of the wide variety of possible crop/animal interactions and the range of related crops and feed sources. Mixed crop/livestock systems are categorized as part of intensive systems in this book but may also be considered a third type of system.

There are good reasons why each of the systems described in Table 11.1 evolved as they did. For example, the Asian systems are adapted to areas of relatively high population density with generally poor soils and intensive use of available resources. By-products such as those from coconuts, marine life or crop production are used in a variety of ways (Winrock International, 1982a). There are, of course, differences between countries and cultures, but as McDowell and Hildebrand point out, "to know and understand the interaction of the system in the coastal area of southern Luzan in the Philippines is to feel familiar with it wherever it is found" (1980, p. 11).

Once a system's orientation is adopted, the analyst subconsciously begins comparing and contrasting, looking for similarities and for differences, as Camoens (1985) did, to include interactions, constraints, strengths, adoption of new technology and intervention strategies. An example for two of the ten Asian systems he documented are provided in Table 11.2. Types of animals are related to food availability or crops and forages with soil fertility (Hart and McDowell, 1985). Animals are identified as a means to convert otherwise low or nonvalue feedstuffs into a marketable or more valuable product with the question being, how, why, when, where, how much, in what way and to what extent? The analyst wants to know why not, as well as why, for the description of livestock systems is just an introductory part to determine ways in which both livestock producers and animal product consumers can be better off (Gartner, 1984). This problem is compounded when the systems are found on small farms, especially those operating at the subsistence level (Winrock International, 1982b).

Table 11.1. Prevailing systems of agriculture on small farms, main regions of use, major crops and animal species, and feed sources for animals of Asia

Farming System	Major Crops	Major Animals	Main Regions	Feed Sources
Coastal fishing and farming complexes, livestock relatively important	Coconuts, cacao, rice, cassava	Swine	P, T	Coconut by-products, rice bran
		Ducks	TW, T, M, P, I	Marine products, rice bran
		Cattle, goats	SL, P, M, I	Pastured with coconuts
Low elevation, intensive vegetable and swine, livestock important	Vegetables	Swine	C, TW, HK	Sweet potato residue, rice bran, fermented residue from vegetable crops
		Ducks	HK	Crop residues, imported feeds
		Swine, fish	TW,M	Crop residues, rice bran

continued

Table 11.1. Prevailing systems of agriculture on small farms, main regions of use, major crops and animal species, and feed sources for animals of Asia (continued)

Farming System	Major Crops	Major Animals	Main Regions	Feed Sources
Highland vegetables and mixed cropping (intensive), livestock important	Vegetables, sugar cane, sweet potatoes	Buffalo, cattle	P, T	Crop residues, rice bran, cut forage, sugar-cane tops
	Vegetables	Sheep, goats	I	
		Swine	P	Crop residues, waste vegetables
	Rice	Cattle, buffalo	Asia	Crop residues
Upland crops of semiarid tropics, livestock important	Maize, sorghum, kenaf, wheat millet, pulses oilseeds, peanuts, etc.	Cattle, buffalo, goats, sheep poultry, swine	IN, T	Bran, oilseed cake, straw, stovers, vines, hulls, hay
Humid uplands, livestock important	Rice, maize, cassava, wheat, kenaf, sorghum, beans	Swine, poultry cattle, buffalo	Asia (> 1,000 mm rain)	Stover, weeds, by-products, sugar cane tops
	Sugar cane	Cattle, buffalo	T, P, I	Sugar cane tops, residues

continued

Table 11.1. Prevailing systems of agriculture on small farms, main regions of use, major crops and animal species, and feed sources for animals of Asia (continued)

Farming System	Major Crops	Major Animals	Main Regions[a]	Feed Sources
Lowland rice, intensive livestock	Rice, vegetables, pulses, mungbean, sugar cane	Cattle, buffalo, swine, ducks, fish	Asia	Crop residues, weeds, by-products, sugar cane tops
Multistory (perennial mixtures), livestock some importance	Coconuts, cassava, bananas, mangoes, coffee	Cattle, goats sheep	P, IN	Cut and carry feeds from croplands
	Pineapple	Cattle	P, I	Crop residues, by-products
Tree crops (mixed orchard and rubber), livestock some importance	Orchard, trees, rubber, oil palm	Cattle, goats, swine	P, M South T	Grazing or cut and carry
Swidden, livestock important	Maize, rice, beans, peanuts, vegetables	Swine, poultry, goats, sheep	Asia	Animals scavenge
Animal based	Fodder crops	Cattle, buffalo, goats, sheep	I, M, IN	Cut and carry fodder, crop residue

Source: McDowell and Hildebrand (1980).
[a]C: China; HK: Hong Kong; IN: India; I:Indonesia; M: Malaysia; P: Philippines; SL: Sri Lanka; Tw: Taiwan; T: Thailand.

Table 11.2. Two of ten major Asian livestock production and management systems

Item	Swidden	Shifting
Characterized by	Jungle dwelling, generally along river. Humid tropics.	Move periodically and return to the same spot every 10-15 years. Asia wide.
Contribution of livestock	High: provides most of animal protein requirements, manure, fuel, mainly from game and wild fowl.	High: provides most of animal protein requirements, manure, fuel, capital accumulation.
Influence of		
Climate	Large	Large
Culture	Strong, generally live as rival tribal groups, avoid strangers.	Strong, generally live and move in family or extended families.
Resources		
On-farm	Labor: tribal and family; Land: none; Livestock: chickens, pigs; Implements: rudimentary; Capital: none.	Labor: family; Land: none; Livestock: chickens, pigs, goats; Implements: some improved tools; Capital: none.
Ex-farm	Forest products, wild game, fish.	Forest products, wild game, fish.
Level of Technology	Generally ignore any technological changes.	Generally aware of technological changes but financial capability to purchase these is limited; may obtain by barter or emulation.

continued

Table 11.2. Two of ten major Asian livestock production and management systems (continued)

Item	Swidden	Shifting
Linkages Public Sector	Generally avoid this sector.	Some contact but largely ignored because of inaccesibility.
Private Sector	Intertribal, some trading.	Between households and with neighboring villages.
Outputs Crops	Root crops, rice, maize, vegetables, all home consumed or bartered.	Rice, maize, root crops, fruits, forest products.
Livestock	Chicken, pigs, manure, fuel, wild game.	Chicken, pigs, manure, fuel, wild game.
Disposal of Products Private Sector	None: home consumed and bartered.	Self-consumed, bartered and traded.
Public Sector	None	None
Interactions	Only within and between tribal groups and friendly villagers. None with formal institutions, infrastructure.	Between neighboring family units and sedentary villages. Some with formal institutions. Use developed infrastructure when unavoidable.

continued

Table 11.2. Two of ten major Asian livestock production and management systems (continued)

Item	Swidden	Shifting
Constraints	Extensive land clearing deprives them of forest products and game. Labor intensive; not accessible to factors influencing change.	Depletion of soil fertility. Reduction of available land because of agricultural development and forestry. Generally resist official interference with their established customs and tradition. Labor intensive.
Strengths	Virtually independent existence and do not utilize subsidies. Capital low.	Their mobility permits wider utilization of natural resources. Some cash generation due to sale of surpluses. Capital low.
Adoption of New Technology	Only reluctantly: they trust their existing, proven technology.	Will adopt new technology if shown to be superior. Financial statis may limit greater modernization.
Intervention Strategy	Much more understanding is required of their perceptions for livelihood and improvement. Locating new settlements near their habitat might induce change.	They are amenable to settlement if their old customs of hunting and fishing are not denied. If the preservation of their tribal customs is assured, they can be grouped into production units on condition that their produce provides regular cash incomes. Generally, provided transportation and markets, they can be persuaded to accept new development strategies.

Source: Adapted from Camoens (1985).

Swine Systems and Economics

There are two major pig production systems in most parts of the world, farrowing and finishing. The first one is production of feeder pigs (piglets), while the second one is feeding them to slaughter weight. In some cases there is a readily definable intermediate growing stage. However, the growing stage is usually incorporated as part of either the farrowing or finishing operation. The approach in this section is to provide examples of several different types and sizes of swine systems in the Peoples Republic of China (PRC) to explain how technical and economic analysis can be combined.

Introduction to PRC Systems

Chinese pig systems can be divided into three regions, the North China Plain, the northeast and the northwest, and the south. The North China Plain includes the area around Beijing and consequently data are from there. The northeast and northwest encompass wide areas, and thus data are averages subject to moderate regional differences. The south is typified by Chengdu, in Sichuan Province. However, the parameters and budgets are applicable to all of southern China with relatively little modification.

A word of caution is, of course, in order at this point as there clearly are many differences between operations. Furthermore, prices fluctuate greatly from year to year as well as seasonally. For example, all data are for August 1986, but at that time feed prices were very high in China. Consequently, care must be taken when interpreting the data as conditions can, and do, change rapidly. This is one reason considerable emphasis is placed on provision of input data so that alternative scenarios could be developed for different situations. A key point is to note how the data are presented and analyzed so that, with relatively little adjustment, similar analyses can be carried out anywhere in the world.

Feeder-Pig Production

There are three general types of farrowing operations in China, those found in suburban areas, those at some distance from metropolitan centers (titled "country"), and those in mountainous regions. In addition, there are three main types within each geographical area; households raising 1 or 2 native sows (called "native"), households with 1 or 2 crossbred sows (called "crossbred"), and state farms. There are a few specialized households with 5-20 sows, but the number is quite limited, so this system is not included.

Production parameters for the suburban farrowing operations, provided in Table 11.3, are based on 1 sow in the small native and crossbred households and 50 sows on the state farms. Similar analyses are developed for country and mountain systems elsewhere (Liu, Simpson, Xiong and Kojima, 1987).

Table 11.3. Production parameters for suburban Chinese feeder-pig-producing operations, 1986

Parameter	Units	North China Plain Small Cross	North China Plain States	Northeast & Northwest Native	Northeast & Northwest Small Cross	Northeast & Northwest State	South Native	South Small Cross	South State
Number of sows	no	1	50	1	5	50	1	1	50
Sow age, first service	mo	7.0	7.5	7.0	7.0	7.5	6.0	7.0	7.5
Sow age at disposal	mo	18	45	48	48	45	54	48	45
Sow weight mature	kg	150	170	130	150	170	130	140	150
Farrowings per year	no	1.7	1.8	1.5	1.7	1.9	1.6	1.7	1.9
Total farrowing, life of sow	no	5.8	5.6	5.1	5.8	5.9	6.4	5.8	5.9
Replacement rate	pct	25	27	25	25	27	22	25	27
Piglets									
Survival rate	pct	85	90	80	80	80	85	85	90
Weaning rate	pct	75	85	70	80	80	85	85	93
Growing rate	pct	90	0	90	90	0	0	0	0
Born	no	11.9	8.0	12.0	9.2	7.3	12.2	9.8	7.5
Survive	no	10.1	7.2	9.6	7.4	5.9	10.4	8.4	6.8
Weaned	no	7.6	6.1	6.7	5.9	4.7	8.8	7.1	6.3
Sold	no	6.8	6.1	6.0	5.3	4.7	8.8	7.1	6.3
Total weaned per year	no	13.0	11.0	10.0	10.0	9.0	14.0	12.0	12.0
Total weaned, sow's life	no	44	34	34	34	28	56	41	37

continued

Table 11.3. Production parameters for suburban Chinese feeder-pig-producing operations, 1986 (continued)

Parameter	Units	North China Plain		Northeast & Northwest			South		
		Small Cross	States	Native	Small Cross	State	Native	Small Cross	State
Piglets (cont.)									
Weaning age	days	60	60	60	60	60	60	60	60
Weaning weight	kg	8	12	8	9	11	7	8	10
Selling age	days	90	60	90	90	60	60	60	60
Selling weight	kg	12.5	12.0	10	12	11	7.0	8.0	10.0
Average no. of boars	no	0	2	0	0	3	0	0	0
Useful life of boars	mo	--	54	--	--	54	--	--	--
Death rate sow head	pct	2	5	5	5	5	5	5	5
Daily concentrate intake									
Gestation	kg	1	1	1	1	1	.5	1.0	1.0
Lactation	kg	2	2.5	2	2	25	2	2	2.5
Daily other feed intake									
Gestation	kg	3	4	3	3	4	2.5	3	4
Lactation	kg	3	4	3	5	4	3	3	4
Piglet feed intake, birth to selling weight	kg	15	2.5	15	15	2.5	2.5	2.5	2.5

Source: Liu, Simpson, Xiong and Kojima (1987).

Sow mature weights vary considerably, from about 130 kg in native breeds to the 170 kg crossbred sows found on most state farms. Sows are usually culled at a very late age, about 60 months in mountain areas and 45-48 months in the other regions. The number of farrowings also varies greatly, from an average of 1.5 annually in most small northeastern and northwestern households to 1.9 on many state farms. Consequently, because the number of weaned feeder pigs is different between systems, the total weaned per sow also varies greatly, from just 28 on northeastern state farms to 56 in southern households with native sows.

Virtually all farrowing operations wean at 60 days. The small ones will sell their feeder pigs at that time while state farms transfer them to finishing operations. Considerable variation exists in selling weights, from 5 kg in southern native operations to 12.5 kg by small crossbred owners in the North China Plain. Death losses also vary considerably, from 35 percent between survival at birth until weaning in northeastern and northwestern households with native sows to just 7 percent on suburban state farms. In contrast, the average selling weight of feeder pigs in Japan is 44.1-47.7 kg at 100 days of age, while the number is 21.3-24.0 kg in the northern United States at about 60 days of age. The sale weight at 60 days in China varies between 6 and 12 kg at 60 days. Even those selling at 90 days only average 10-12 kg.

Finished-Hog Production

The number of feeder pigs finished to slaughter weight varies from 1 head by small native pig producers to 500 head or more on North China Plain state farms. Specialized households finishing crossbreds raise between 7 and 20 head depending on the region.

Fattening period, and thus age at slaughter, ranges widely in China depending on the size and type of operation. To a lesser extent, this is also true of the small crossbred operations because small producers usually target production to correspond with peak marketing periods, such as festivals, when prices are higher than normal. In addition, many producers have access to abundant crop residues or other forage type feedstuffs during certain periods of the year. Thus, producers deliberately slow down production, especially in the growing phase. As a consequence, finished hogs in mountain areas average 420 days of age at slaughter while those on suburban state farms will be half that, 210 days (Table 11.4). Almost all Chinese hogs are sold at much lighter weight, about 80-90 kg.

The average daily gain in China varies greatly, from 0.21 kg by native hogs in the southern mountain areas to 0.44 kg on most specialized households and state farms. In general, death loss is 5-10 percent in China, except on North China Plain suburban operations where it varies between 2 and 3 percent. Feed conversion ratios are hard to compare because most Chinese producers frequently feed by-products as well as mixed feed.

Table 11.4.   Production parameters for Chinese pig
               finishing operations, 1986

| | | Suburb | | | |
| | | Small | | | |
Parameter	Units	Native	Cross	Spec.	State
		North China Plain			
Initial number	hd	--	4	20	500
Initial age	days	--	60	60	60
Initial weight	kg	--	10	10	12
Fattening period	days	--	210	180	150
Age of slaughter	days	--	290	240	210
Weight at slaughter	kg	--	90	90	90
Total gain	kg	--	80	80	80
Average daily gain	kg	--	.38	.44	.53
Feed conversion	ratio				3.6
Death loss	pct		2	3	3
		Northeast and Northwest			
Initial number	hd	--	3	10	250
Initial age	days	--	60	60	60
Initial weight	kg	--	10	10	12
Fattening period	days	--	210	180	150
Age of slaughter	days	--	290	240	210
Weight at slaughter	kg	--	90	90	90
Total gain	kg	--	80	80	80
Average daily gain	kg	--	.38	.44	.53
Feed conversion	ratio	--			
Death loss	pct	--	10	5	5
		South			
Initial number	hd	--	2	20	--
Initial age	days	--	60	60	--
Initial weight	kg	--	8	10	--
Fattening period	days	--	180	180	--
Age at slaughter	days	--	240	240	--
Weight at slaughter	kg	--	80	90	--
Total gain	kg	--	72	80	--
Average daily gain	kg	--	.40	.44	--
Feed conversion	ratio	--			--
Death loss	pct	--	10	5	--

Source: Liu, Simpson, Xiong and Kojima, 1987.

Feed Price and Ration Analysis

Relative values of Chinese feedstuffs for August 1986 for Chengdu based on priced corn are given in Table 11.5. A swine ration analysis and relative value computer program from the Iowa State University Cooperative Extension Service was modified and used to analyze the ration costs and the budgets presented in other sections. The calculated relative values are based on the nutrient content of the feed for energy, lysine and phosphorus. These values, in turn, are based on the prices and composition of corn, soybean meal and dicalcium phosphate.

Review of the relative value tables shows that considerable disparities exist between feedstuffs thus indicating that prices are not being set according to relative feeding values. For example, the Chengdu data, based on a free-market corn price of $13.52 per 100 kg (the government price in August 1986 was $10.27) show that the relative value of free-market-priced wheat bran was $12.39 per 100 kg. (From this point all feed related values are expressed in terms of 100 kg unless otherwise indicated). The actual price was $10.81, which means that a $1.58 advantage existed in favor of wheat bran over corn. Rapeseed cake had an $8.12 advantage, while peanut cake had a $3.08 disadvantage.

Analysis of relative value data for Japan and the United States indicates there is very little, and in most instances no, relative advantage or disadvantage between feedstuffs because feedstuffs are widely traded daily, and as soon as a relative advantage appears for a feedstuff demand for it immediately increases, thereby driving up the price. In contrast, there are no published price data for feedstuffs in China and virtually no use made of relative value analysis. Only very limited use is made of least-cost feed formulation. Equally as important is the great fluctuation in feedstuff availability and quality. The same holds true for most developing countries.

A typical finishing ration for hogs weighing 60-90 kg containing 12.34 percent protein and 1366 Kcal of energy ration recommended by some researchers in Chengdu would have cost $12.73 per 100 kg in August 1986, assuming free-market-priced corn (Table 11.6). A least-cost ration computer program using linear programming was used to choose a similarly balanced ration. The result is a ration based mainly on dried sweet potatoes and rice bran (about 80 percent) that would have cost $6.67, or just a little more than one-half that of the recommended ration, which has 62 percent corn.

A comparison is useful because it clearly demonstrates the extent to which the cost of mixed feed can be decreased by use of least-cost formulation in certain situations. It is clear that if price information and the least-cost ration formulation technology were widely known, the extreme price differences would disappear. Nevertheless, the comparison provides insight into possibilities for macro-level analysis about which crops to recommend in order to reduce feed costs. Formulation of various rations also serves as a key cost factor in carrying out economic

Table 11.5. Actual price, relative feeding values and
economic advantage of swine feedstuffs based on
free-market-priced corn, Chengdu, August 1986

Ingredients	Actual	"Relative Values"		$1/100kg
	$/100kg	$/kg	$/100kg	Adv.
Azolla, wet	0.27	0.004	0.42	0.15
Barley	14.6	0.129	12.93	-1.67
Blood meal	48.54	0.344	34.41	-14.23
Bone meal	0	0.179	17.86	0.00
Brewers grains, wet	2.71	0.030	3.02	0.31
Broken rice	9.19	0.140	13.99	4.80
Cabbage	1.08	0.012	1.16	0.08
Corn, average	13.52	0.135	13.52	0.0
Corn gluten	2.71	0.173	17.27	14.56
Cottonseed meal	7.56	0.145	14.47	6.91
Dicalcium phosphate	20.52	0.205	20.52	.00
Dry silkworm powder	32.14	0.258	25.82	-6.62
Fishmeal, domestic	43.22	0.226	22.61	-20.61
Fishmeal, imported	54.04	0.313	31.31	-22.73
Horse beans, dry	18.92	0.168	16.78	-2.14
Peanut cake	19.45	0.164	16.37	-3.08
Peas, dry	20.53	0.185	18.45	-2.08
Rapeseed cake	8.64	0.168	16.76	8.12
Rice bran	5.04	0.143	14.30	8.90
Rice bran, defatted	4.32	0.129	12.92	8.60
Sesame cake	17.30	0.159	15.90	-1.40
Sorghum, gov't.	10.81	0.130	13.00	2.19
Sorghum, free	16.22	0.130	13.00	-3.22
Soybean meal	20.54	0.205	20.54	.00
Sweet potato, dry	5.41	0.131	13.08	7.67
Sweet potato vines, F	1.08	0.010	1.02	-0.06
Sweet potato vines, D	0	0.070	6.96	0.00
Tofu (beancurd) waste	2.70	0.014	1.45	-1.25
Water hyacinth	0	0.005	0.53	0.00
Wheat bran, gov't.	0	0.124	12.39	0.00
Wheat bran, free	10.81	0.124	12.39	1.58

Source: Liu, Simpson, Xiong and Kojima, 1987.

Table 11.6. Comparison of recommended and least-cost
finishing-pig rations, Chengdu, August 1986

Ingredient	Pro-tein	Price	Price advantage	Recom-mended	Least cost
	-Pct-	- - $/100 kg - -		- - Pct - -	
			Composition		
Corn, free market	8.60	13.52	0.00	62.10	
Cottonseed meal	33.80	7.56	6.91		10.000
Dicalcium phosphate	0.00	20.52	-6.62	1.20	1.517
Dry silkworm powder	53.90	32.44	-6.62	1.00	--
Lycine	0.00	378.00	0.00	.28	
Methionine	0.00	324.00	0.00	.12	
Peanut meal	43.90	19.45	-3.08		1.937
Premix	77.00	77.00	0.00		.540
Rapeseed meal	36.40	8.64	8.12	8.00	5.000
Rice bran	12.10	5.40	8.90	24.00	38.194
Salt	0	3.24	0.00	.30	
Sweet potato, dry	3.90	5.41	7.67		42.812
Wheat bran,free mkt.	14.40	10.81	1.58	3.00	
			Analysis		
Total ingredients (pct)				100.00	100.00
Ration cost ($/100 kg)				12.73	6.67
Protein (pct)				12.13	12.34
Energy (k cal)				1366	1304

Source: Liu, Simpson, Xiong and Kojima, 1987.

analyses of the swine systems for which production parameters were presented earlier.

Economic Analysis, Current Systems in China

Input and output values for suburban pig-producing systems are provided in Table 11.7. The results were calculated by using a computer-based cost and returns program that employs 39 variables. Included are 15 production variables whose parameters are among those in Table 11.3.

There is a negative return to labor and management in every system, ranging from $-0.85 per feeder pig sold or transferred to finishing operations for state farm operations in the north to a positive return of $0.07 by small native pig producers in the south. In virtually all cases, producers with 1 native sow lose the least money while state farms suffer

Table 11.7. Input and output values for suburban China feeder-pig production, August 1986

Item	Units	North China/Plain Small		Northeast & Northwest Small			South Small		
		Cross	State	Native	Cross	State	Native	Cross	State
Number of sows	no	1	50	1	1	50	1	1	50
Selling price feeder pigs	$/100kg	9.12	8.75	7.57	9.08	8.32	4.92	5.62	7.03
Investment in facility per sow	$/hd	28.04	42.76	24.65	24.65	42.77	3.13	3.13	8.29
Concentrated per price	$/100kg	11.92	11.23	11.92	11.92	11.23	11.92	11.92	11.23
Other feed price	$/100kg	0.07	0.54	1.08	1.08	1.08	0.008	0.008	0.011
Labor per sow	hr/yr	3.2	468	312	312	468	312	312	468
Cost per pig sold									
Feed	$/hd	9.04	5.81	12.67	12.85	11.64	4.00	6.07	7.33
Other variable costs	$/hd	4.45	9.25	5.84	5.90	8.70	3.56	4.24	6.15
Total variable costs	$/hd	13.49	15.06	18.51	18.75	20.34	7.56	10.31	13.48
Total fixed costs	$/hd	2.42	2.83	2.80	2.81	4.87	0.23	0.22	0.66
Total cost	$/hd	15.91	17.95	21.31	21.56	25.22	7.79	10.58	14.14
Income per pig sold	$/hd	9.62	10.72	9.94	11.73	11.66	6.22	7.42	9.19
Profit & return to mgt.	$/hd	-6.28	-7.23	-11.37	-9.83	-13.56	-1.58	-3.16	-4.95
Return to labor & mgt.	$/hr	-0.09	-0.85	-0.18	-0.14	-0.36	0.07	0.02	-0.11

Source: Liu, Simpson, Xiong and Kojima, 1987.

the greatest loss. The main reason is although state farms have the largest income per pig sold and also have larger numbers of pigs weaned per sow per year, feed costs and labor costs are much higher.

The profit and return-to-management results include a charge for labor. But small producers do not pay themselves a wage; in general, they consider residual income after other costs are covered to be their income. Analysis in this way shows that almost all producers have a negative return per hour. For example, owners or family members of households with crossbred sows in the North China Plain spend about 312 hours per year with their 1 sow. This time includes obtaining feed and giving it to the sow and piglets, caring for them, selling the feeder pigs, and so on. The result is a loss of about $0.09 for each hour spent. Even if no charge were made for fixed costs, such as money tied up in the facility and depreciation of it, there would still be a loss.

The major reason for negative returns is relatively high cost of feedstuffs. Thus, even though owners of only 1 or 2 sows will feed some of the sows table scraps and by-product feeds, owners still must provide about 2 kg of mixed feed or grain during lactation and 1 kg during gestation. In contrast, operations only feeding concentrate give about 2.5 kg during gestation and 5 kg during lactation. Thus, even though so-called "backyard" operations with 1 or 2 sows are thought to have little or no costs, in reality these operations do have considerable outlay.

Feeder-pig production continues despite the calculated losses because there is virtually no record-keeping and small producers think they have little or no production cost. In some cases, producers do have access to low-cost feedstuffs such as garbage from hotels or other institutions, from industrial or crop by-products and from low-priced mixed feeds. State farms produce because they need large numbers of feeder pigs for their finishing operations. In addition, they consider themselves to have a social mission so that, while there may be losses in feeder-pig production, those losses are covered from profits in other state farm industries.

Table 11.8 contains input and output values for suburban finished hog production. In contrast to feeder-pig production, many finished-hog producers make respectable profits and return to management. The advantage of size and technology are apparent for state farms, most of which have about 500 head, make the highest profits, and are followed closely by specialized households with 20 or so head. Small producers finishing just 2-4 head of crossbreds lose money, except in mountain areas where hogs graze part of the year.

Economic Analysis, Improved Systems in China

Review of the production parameters and economic analyses just carried out indicates that production efficiency can be improved and that production costs thus could be reduced. One way is to modify existing systems in such a way that they still fit within the realities of China's production possibilities situation. Much attention has been given to small scale operations with particular emphasis on the use of crossbred sows.

Table 11.8. Input and output values per suburban China finished-hog production, August 1986

Item	Units	North China Plain			South	
		Small Cross	Spec-ialized	State	Small Cross	Spec-ialized
No. of pigs purchased	hd	4	20	500	2	20
Selling price	$/100kg	48.64	48.64	48.64	47.60	53.50
Investment in facility	$/hd	8.11	13.51	20.62	6.76	13.51
Labor per head	hr	52	36	14	52	36
Concentrate feed price	$/100kg	11.00	11.00	10.50	10.00	10.00
Other feed price	$/100kg	0.07	0.07	0.07	1.08	1.08
Cost per pig sold						
Feeder pig	$/hd	7.33	7.33	8.76	5.62	7.03
Feed	$/hd	35.64	31.18	28.77	31.03	29.83
Other variable costs	$/hd	8.86	8.93	3.78	8.50	8.90
Minus manure sales	$/hd	3.41	2.92	2.43	2.92	2.92
Total variable costs	$/hd	48.42	44.52	38.88	42.23	42.84
Total fixed costs	$/hd	0.31	0.45	0.57	0.22	0.15
Total cost	$/hd	48.73	44.96	39.45	42.45	43.29
Sales price per head	$/hd	43.78	43.78	43.78	38.08	48.15
Profit & return to mgt. (inc. death loss)	$/hd	-4.71	-1.13	4.11	-4.15	4.62
Return to labor & mgt.	$/hr	0.05	0.18	0.46	0.06	0.34

Source: Liu, Simpson, Xiong and Kojima, 1987.

Consequently, that is one of the two systems analyzed. The other system is a large one.

Production parameters for the improved feeder-pig production system given in Table 11.9 are based on good management and excellent breeding stock. But investment in facilities is relatively low (part of the realities of China). Nevertheless, it represents the best possible operations, some of which are now found in selected areas. There will be small regional differences due to climate, feed cost differences, and so on, but in general the system has relevance to most of China for analytical purposes.

The economic analysis reveals that even under the most optimal conditions set forth in the improved system, a feeder-pig producer would lose $2.43 per feeder pig produced (Table 11.10). In addition, even though labor use is the lowest of all systems there would be a negative return of $0.03 per hour.

Production parameters for the improved finishing operation are given in Table 11.11. The feed conservation is 3.3, and the average daily gain is 0.85. Hogs are sold at 90 kg.

The economic analysis shows that the specialized producer would earn $4.32 per pig produced and have a $0.36 return per hour. The large producer with 500 head and about the same production parameters but benefiting from economies of size would earn $5.63 per pig fattened and $0.56 per hour.

Development Strategies

China's swine industry is changing rapidly in concert with the transition taking place in the general economy. Each year more elements of a free-market system are added, and the swine industry is one indication that the new system is working well. On the demand side, consumers through the free market price system are effectively expressing their desire for leaner pork and a greater proportion of their pork purchases as meat rather than fat. On the supply side, a shift is rapidly taking place to produce lean hogs, thanks to government action in promoting this activity.

A key element in evaluation of China's swine industry is having national-level goals or objectives on which recommendations can be based. To date, no specific set of objectives have been issued by the government about the swine industry. But, it can be postulated that the following objectives are reasonable and would likely be embraced by planners. These objectives can serve as useful guides for a final analysis of the swine industry and as points to which attention should be given. They are

1. Decrease price of swine products to consumers
2. Increase income to producers
3. Assure a steady supply of pork and other swine products
4. Improve pork product quality in response to consumer desires

Table 11.9.   Production parameters for improved feeder-pig production

Production Parameter	Units	North China Plain Small	State
Number of sows	no	3	50
Sow age, first service	mo	7.0	7.5
Sow age at disposal	kg	24	24
Sow weight, mature	no	200	200
Farrowing per year	no	22	22
Total farrowing, life of sow	no		
Replacement rate	pct	50	50
Piglets			
Survival rate	pct	95	95
Weaning rate	pct	93	93
Growing rate	pct	96	96
Born	no		
Survive	no		
Weaned	no	9.0	9.0
Sold	no		
Total weaned per year	no		
Total weaned, sow life	no		
Weaning age	days	30	30
Weaning weight	kg	7.0	7.0
Selling age	days	60	60
Selling weight	kg	20	20
Average no. of boars	no	0	0
Useful life of boars	mo	0	24
Death rate sow herd	pct	5	5
Daily concentrate intake			
Gestation	kg	2.0	2.0
Lactation	kg	3.5	3.5
Daily other feed intake			
Gestation	kg	7.0	7.0
Lactation	kg	7.0	7.0
Piglet feed intake, birth to selling weight	kg	20	20

Source: Lin, Simpson, Xiong and Kojima, 1987.

Table 11.10. Input and output values for improved China feeder-pig and finished-hog production, August 1986

	Units	Feeder pig production		Hog finishing	
		Small	Large	Small	Large
Number of sows	hd	3	50	--	--
Number of pigs purchased	hd	--	--	20	500
Selling price					
Feeder pigs	$/100kg	16.22	16.22	--	--
Finished hogs	$/100kg	--	--	62.16	62.16
Investment in facility	$/hd	23.59	16.35	1.08	1.08
Concentrate feed price	$/100kg	11.92	11.92	11.50	11.50
Other feed price	$/100kg	1.08	1.08	0.00	0.00
Labor per hog sold	hr	--	--	21.6	13.8
Labor per sow	hr	243	390	--	--
Cost per pig sold					
Feeder pig	$/hd	--	--	16.22	16.22
Feed	$/hd	13.80	13.80	26.57	26.57
Other variable costs	$/hd			10.34	8.96
Minus manure sales[a]	$/hd	--	--	1.75	1.75
Total variable costs	$/hd	18.63	19.65	51.38	49.99
Total fixed costs	$/hd	1.25	0.87	0.02	0.02
Total cost	$/hd	19.89	20.52	51.40	50.01
Sale price per head	$/hd	--	--	55.94	55.94
Income per pig sold	$/hd	17.46	17.46	--	--
Profit & return to mgt.(inc death loss)	$/hd	-2.43	-3.06	4.32	5.63
Return to labor & mgt.	$/hr	-0.03	-0.42	0.36	0.56

Source: Liu, Simpson, Xiong and Kojima, 1987.
[a]Included is income for feeder-pig production.

Overall, it is apparent that virtually every country's major objective is to improve the industry's efficiency. Analysis of the LDC's swine industries, including production parameters and economic budgets as well as relative values of feedstuffs, indicates that considerable improvement can be made in efficiency and thus in reducing production cost.

Table 11.11. Production parameters for improved Chinese
pig-finishing operations

Parameters	Unit	Small	Large
Initial number	hd	20	50
Initial age	days	60	60
Initial weight	kg	20	20
Fattening weight	days	108	108
Age at slaughter	days	168	168
Weight of slaughter	kg	90	90
Total gain	kg	70	70
Average daily gain	kg	.65	.65
Feed conversion	ratio	3.3	3.3
Death loss	pct	5	5

Source: Liu, Simpson, Xiong and Kojima, 1987.

A first step in reducing production cost is to develop grading standards for swine. This does not mean immediately instituting a grading system in which all hogs are sold on a graded basis. Rather, the grades should first be set up as production targets. In effect, it is first necessary to determine what will be produced. Only then is it possible to determine how swine production can be improved.

The next step is to decide on one or two breed types for various regions of the country. For example, there may be 10 different production conditions determined for a country. In that case, there might be 5 or 10 breeds or breed types. There is sufficient experience in cross-breeding that it is now possible to determine the optimal breed or cross for each region.

The analysis provided in this chapter on relative feeding values shows there is a very wide range in price advantage and disadvantage of feedstuffs in China. The same holds true in any developing country. In most economically developed country's feedstuffs relative values only vary 1-2 percent from each other. But the analysis provided indicates the price disparities in China are often plus or minus 50-75 percent of each other. In effect, not only are feeds not being formulated on a least cost basis, but there also is a waste of natural resources.

Quality improvement is also critical to more rationalized use of feedstuffs. At present, feedmills in most LDCs as well as China have virtually no control over quality of feedstuffs purchased. Failure to test for quality of feedstuffs purchased leads to sales of feedstuffs that are not only highly variable in quality but also can be dangerous to an animal's health. Equally as important, producer confidence in the use of manufactured feeds, and especially in formulated ones, is eroded.

There are several major areas to which research efforts in swine nutrition in both China and all LDC's should be devoted. A main area is comparison of economic results from a wide variety of feeding systems, management methods and rations. For example, there is great need to understand the physical and economic impact of feeding large amounts of low-priced feedstuffs in a variety of conditions. Another subject is comparison of current feeding practices with recommended ones. The effort there should be on development of data and results that extension-related investigators can use in their development efforts.

Measuring the Value of Labor

Nutrition is closely tied to management. Thus, the two need to be researched together. One major area that needs attention is use of home-produced feedstuffs, and especially of so-called nonconventional feedstuffs such as sweet potato vines, food processing by-products and homegrown feedstuffs like tubers. Producers may utilize these feeds in the belief that they are essentially free because there is no direct monetary outlay in purchase of them. But there is an opportunity cost involved, and that opportunity cost should be measured, appropriate economic analyses carried out and results disseminated to producers.

Producers who are feeding pigs homegrown feeds have harvest time involved. This time could be spent in other activities; thus, the collection should be charged at the rate producers could earn in those alternative activities. In addition, some crops may be grown because producers calculate that part of the production cost should be allocated to the feed. It is true that labor in China has little mobility, and generally there are quite limited alternative activities. But, it is also true that although monetary values may be somewhat difficult to calculate, recognition of opportunity cost is critical to sound recommendations about industry improvement.

One way to relate labor opportunity cost with feedstuffs values is to parametrically change the value of homegrown feedstuffs using a relative value program (like the one described earlier in this chapter) until the price advantage is equal to zero. For example, fresh sweet potato vines were priced at $1.08 per 100 kilos in August 1986 in Chengdu. Sweet potato vines had a price disadvantage of $0.34 relative to government-priced corn and of $0.06 relative to free-market-priced corn. Let us make an analysis related to government-priced corn. Parametrically changing the price of the fresh sweet potato vines until the advantage is zero results in a comparative price of $0.74.

The opportunity cost value can be determined by calculating the time required to harvest, transport and store 100 kg of vines. Suppose it takes 8 hours to do this. Division of the comparative value of $0.74 by 8 means the relative value per hour is $0.09. Furthermore, suppose there is no sweet potato production cost assigned to the vines, and there is no value as mulch. The rational producer would then determine if there were an alternative activity that would provide an income greater that $0.09 per hour. If so, then the most profitable thing would be to purchase another feedstuff for the pigs and to work at the other activity. But if there were no alternative that would provide more income than

$0.09 per hour, the most rational thing would be to use the vines for pig feed.

The No-Cash-In/No-Cash-Out Principle

An important principle in livestock analysis work, especially in LDC's, is that an individual will not likely make an investment or use purchased inputs unless additional cash can be generated to recover that expense. Consider, for instance, the Thai example of raising buffalo covered in an earlier chapter. These animals are used as sources of draft power for home farming activities, the products of which are only sold to a limited extent. Thus, without additional cash income to offset the cash outflow, an investment will not be made.

The no-cash-in/no-cash-out principle is one reason why many smallholders seem to be slow in their adoption patterns. But closer analysis reveals that the difficulty is often not that smallholders do not know about the technique or are not convinced of it, but rather they have a type of cash flow problem. Examples are slow adoption of efficient plows and improved harness and the use of purchased animal health inputs.

The number of management strategies available to small scale livestock owners is extremely limited compared with commercial operations because producers near the subsistence level are less likely to purchase production inputs as well as make large (relatively) investments in fixed facilities, equipment and breeding stock. The net result is that in terms of efficiency and productivity, most livestock owners at or near the subsistence level are already operating at a fairly high level of productivity given the constraints under which they operate. This concept can be thought of in the linear programming framework presented earlier, where cash flow considerations replace the prestige constraint or could become another constraint.

Producers need to become part of a market economy if they are to make livestock-related investments either with their own capital or with borrowed capital. In some cases, such as milk from a dairy operation or wool from sheep, the return will be derived directly from the animals themselves. In other cases, the return will be indirect, such as using buffalo for draft power. Thus, the need for the holistic approach to the entire farming system advocated in the FSR/E methodology.

The no-cash-in/no-cash-out principle is a serious constraint to livestock industry improvement, especially where a substantial proportion of the producers operate at or near a subsistence level. However, many opportunities do exist to help these producers improve their productivity. Government can encourage commercial projects, such as organization of villages to specialize in commodities that will be sold rather than consumed locally. Subsidized harness and tillage equipment can be made available. The critical point is to carry out integrated economic and technical analyses with tests being made on farms.

An important aspect is that the no-cash-in/no-cash-out principle holds at the national as well as individual farmer level. Failure to recognize it has led to many national-level livestock projects that have simply added to a country's external debt problem. For instance, range management projects can be very expensive in terms of foreign exchange requirements but often yield little additional offtake for export. Rather, the benefits are often more welfare related. Thus, ironically, even though any particular project itself be very successful in terms of its own scope of work, that project may make virtually no contribution, either directly or indirectly, to generating foreign exchange earnings to repay the international loan.

CHAPTER 12

ANALYSIS OF NEW TECHNOLOGIES

The objective of this last chapter is to round out the principles developed earlier by discussing them as part of the technological change process. We begin with a methodological discussion, then highlight the multitude of livestock production technologies now available or that will become available by 2000. The concluding section incorporates the principles into a framework for analysis.

The Need for a Vision

Producers, researchers and national-level planners alike must have a vision of what they want to accomplish during the longer term in order to optimize shorter-run goals. In effect, each change analyzed and ultimately initiated should be part of a complete plan to obtain the highest yield from the resources available. An analogy is traveling through a rather narrow tunnel. If a well-developed plan is available and care taken at each turn, the passage will be smooth and relatively effortless. But if only brute force is applied, the passage will result in a series of jerks, stops and bouncing off walls culminating in longer travel time and a tortuous, unpleasant experience. Naturally, despite a well-developed vision, continual adjustment will be necessary about course and speed as new constraints are identified and technologies and resources become available.

The scientific method outlined in Chapter 1 wherein the researcher continually travels back and forth between the world of facts and the world of experimentation and analysis is a never ending circular process characterized by aimless wondering if no vision has been established prior to initiating the process. However, if a clearly defined long-term objective has been set forth, then moving from facts to theories by induction and then to predictions by deduction and finally back to the world of facts by verification simply becomes a process of identifying short-term goals and meeting them within the context of a longer-term plan.

The scientific method, just like the principles and methods set forth earlier, has as much relevance for producers as for researchers. For example, a long-term vision might be development of the most outstanding herd of a certain breed cattle in 10 years subject to certain resource limitations and income restrictions. Once that vision has been established, a time frame with quantifiable periodic goals can be set forth. Enterprise budgeting can be used as a tool to plan and monitor economic progress, while partial budgeting combined with capital budgeting can be used to evaluate new information. Part of the process is continual redefinition of goals in light of new information and technology development. The example provided is for a commercial operation, but the concept holds just as well for even the smallest subsistence-level producer. The successful producers will have a vision, however well articulated, of what they want and how to obtain it. In this light, use of credit, risk taking and technology adoption all become meaningful elements of a composite whole. The bottom line is planning; the strategy is proper application of analytical methods.

It is the process of developing a long-term vision that can so carefully wed the scientific method set forth in Chapter 1 with the FSR/E methodology described in the previous chapter. In other words, although experiment-station research is clearly needed as part of the experimentation and analysis phase, another whole phase--deductions from theories to predictions and returning to the world of facts through verification and finally the whole inductive process--implies continual interaction with producers. That process should not simply be a verbal dialogue but rather must include meaningful cooperative research and analysis from the producer's viewpoint--which translates to added returns to net income.

National goals are typically couched in terminology such as annual economic growth rate, reduction of cost to consumers, and national level output increases. But those objectives are met because producers find it in their self-interest to adopt or reject technologies. In this light, attainment of national-level visions becomes a process of providing and interpreting alternatives rather than pushing production through state planning and ownership of means of production. Technology for livestock production abounds and is continually growing. The problem is to assist producers to evaluate that technology in light of their situations.

The Great Leap in New Technology

A virtual revolution has taken place in livestock production and marketing techniques in much of the world during the past two decades. As a result, the EEC has shifted from being a major beef deficit region to a net exporter. Japan, while continually increasing beef imports, has made great improvements in its production and marketing structure. In the United States, beef production per head of cattle inventory increased from 55 kilos in 1950 to 70 kilos in 1960 and is now about 90 kilos.

Changes during the past quarter century have been spectacular, but they pale in light of potential developments during the next two decades. The purpose of this section is to describe and evaluate, using concepts developed earlier, some of the innovations related to feed production and storage and changes in the beef and milk marketing system. Many of the techniques discussed are now beginning to be applied; others are still in their most basic stages. Together these techniques represent a fascinating array of fruits from modern science with opportunities for income improvement by some producers and greater hardship for others. These are some of the principle characteristics of future world animal agriculture.

History is most important for placing current and future animal agriculture in perspective. A fundamental approach is to organize agricultural change into eras as a means of "feeling" the dynamics taking place. The concept of eras also provides a means to understand the why of processes rather than simply extrapolating from past trends. Most countries are now in what Lu, Cline and Quance (1979) term the era of "science power." Previous eras identified by them for the United States were hand power up until about the Civil War in the 1860s, horse power until about World War I, and mechanical power until about World War II (Figure 12.1).

All countries of the world have gone through or are now engaged in one or more of these eras. This analysis will extend the concept and argue that the era we are now entering should be called the "era of science and knowledge power." The reason is that knowledge, at all levels of an economy, acts as both a constraint and a resource in effective adoption of scientific advances in animal and crop agriculture. Futhermore, as will be shown, there is reason to believe the technology change curve will appear similar to the one in the recent past.

Technology, a human-made phenomenon that can be increased through research and development, has begun to be recognized as a resource because of the complementary relationship among various technologies. Imagination, time lag in adoption and institutional constraints are becoming the principal limits to expanded animal productivity rather than the traditional three: land, labor and capital. Technological changes related to livestock and meat production are occurring so rapidly that even scientists closely associated with each field of endeavor are finding it difficult to assess the implications for commercial application. The objective of the next several sections is to provide an understanding of the range and breadth of new technologies that have been recently developed or that can be expected to have commercial application within the next decade.

Animal-Related Technologies

Numerous genetic-oriented technologies are being developed or have recently been made available. Examples are heat period control to

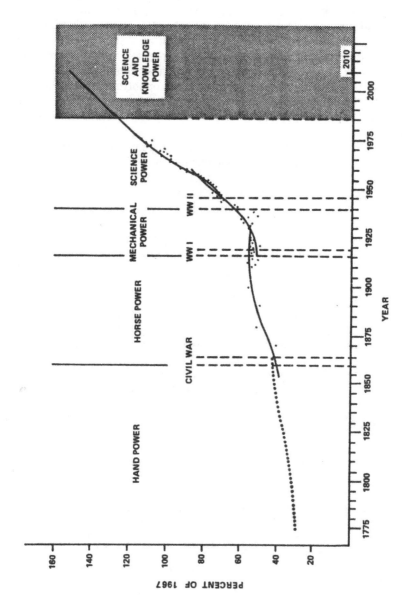

FIGURE 12.1 U.S. AGRICULTURAL PRODUCTIVITY GROWTH -- THE ERA OF SCIENCE & KNOWLEDGE POWER.

improve conception, embryo transfer to reduce time for genetic improvement and/or to obtain multiple calves, and embryo splitting to increase the number of embryos and as a genetics research device using multiple identical births. Great advances have also been made on sexing embryos and freezing them for long-term storage. Even more exciting is the possibility of sexing bull semen. This technique will revolutionize the cattle industry in many parts of the world.

Other genetic technologies include porcine growth hormone to improve growth efficiency, bovine growth hormone as a means to expand milk production from dairy cows, embryo infusion of human growth hormone genes to alter animal genetic makeup, and bred-in parasite resistance. Knowledge in the form of theory and application of genetic principles has led to more sophisticated sire evaluation programs, crossbreeding and development of genetically superior calves.

Animal health improvements include recombinant DNA (termed cDNA by molecular biologists and rDNA by industry) developed vaccines, bovine interferons, electronic mastitis detection, implanted identification tags, use of ivermectins for parasite control, and monoclonal antibodies. Recently, there has been an explosive discovery of new hormones that regulate various physiological functions. The implications in this area are enormous. Indeed, animal health is one of the most rapidly expanding agribusiness areas. Overall, the products and techniques are much further advanced than is adoption or even producer knowledge of them.

The nutrition area encompasses a multitude of new products. Considerable improvement can be expected in current products such as anabolics (growth stimulants) and feed additives. Furthermore, producers have had, and will continue to have, a broad range of new practices and products--what might be called management tools--made available to them. Some recent advances are use of sodium bicarbonate to enhance feed conversion and magnesium to increase milk yield, treatment of straw, hay and crop residues by ammonification and hydrogen peroxide to increase digestibility and energy value, and recycling of wastes from animals. One example for medium to large feedlots with slated floors is recyclable plastic pellets to provide roughage in animal feeds. Other nutritional advances include supplement selection procedures, rumen-regulating drugs (rumen metabolites), microflora use, expanded by-product use, computerized feeding control, and milking machine equipment linked to computers for ration formulation.

Processing and marketing improvements abound. Examples are on-farm extraction of water from milk to reduce transport costs, irradiation of meat to expand storage life, and use of robots in meat processing plants to reduce costs. Restructured meats made from low priced cuts to resemble high price cuts, electrical stimulation and mechanical tenderization of meat are now gaining wide acceptance. New techniques now being researched and tested on a pilot basis are hot carcass processing, cooking and portion packaging immediately after slaughter, and grading of carcasses by computer.

Crop- and Forage-Production Technologies

Animal agriculture of course is dependent on crop and forage techniques. Thus, assessment of what is ahead in animal agriculture is highly related to understanding changes in feedstuffs production.

Biotechnology opens the door for rapid increases in plant improvement through recovery of desirable plant genotypes from tissue culture, protoplast fusion and cloning. Recombinant DNA facilitates the direct manipulation of an organism's genetic material to produce offsprings with desired characteristics. Although use of rDNA is still in the early stages of development, rDNA will be widely applied in the 1990s to agriculture.

Plant breeding work includes increased photosynthetic efficiency, seeds produced by genetic engineering, seed treatment, plant growth regulators, salt tolerance and introduction of polyacetylenes and nitrogen-fixing organisms.

Some of the more exciting agronomic practices include no-till (zero tillage), custom prescribed tillage, forage quality improvement through management, hydroponics, multiple cropping, intercropping, and improved water conservation practices.

Mechanization is an important new area to reduce costs of animal feedstuffs. Innovations include solar bin buildings, controlled traffic, automatic guidance mechanisms, improved engine, draft and tractive efficiency, integrated control harvesting, and computer-controlled spot application of herbicides on weeds. Many of these technologies will have as much impact on very small plots typical of countries in Asia as on the largest farms in North America.

Data Management and Coordination Techniques

Perhaps the most important aspect in the new scientific revolution for agriculture as a whole is data management, and microcomputers are the key hardware. A virtual information explosion has taken place in the past five years due to the greatly expanded use of computers for data analysis and dissemination of information. Computer ownership has also led to improved management practices, a type of psychological synergistic effect.

As an example of computer age data management, some farms now have microcomputers linked to passive transponders that permit automatic milk yield recording, individual cow feed formulation and continuous health monitoring. Coordination in production and marketing will also receive more emphasis as data management constraints are overcome.

Review of animal and crop technological change leads to a conclusion that management is the central variable for effective adoption of them. This conclusion is reinforced by recognition that, the "product life" of each technology and product is becoming shorter as greater knowledge about basic processes is acquired. The net result is that we cannot project with any accuracy, much less predict, the impact of most technologies because they are continually changing. However, when taken as a whole--and in combination with the synergistic effects--the impact will be enormous.

Potential Impact on Production

The technologies just described are impressive as to scope and potential for improving productivity and reducing costs. But it is obvious that the adoption of such technologies, as well as the multitude of technologies now available but not being used, vary from country to country and even from farm to farm within a district. Some reasons for differences are management time and capabilities, land characteristics, production complementarily, tradition and social conditions, location, level of effective demand (purchasing power) for animal-related products, input/output price relationships, and government policies. Productivity analysis is a principal way to combine all these variables for research purposes.

Productivity is a measure of the technological efficiency with which resources are converted to commodities and services. There are two types of productivity: partial productivity and total-factor productivity. Partial productivity can be measured by ratios of output to a single input in which both the numerator and the denominator are measured either in physical units or constant money values. Total-factor productivity presumably is most useful, but difficulties abound in actual measurement due to disparate quantities of inputs. One way economists have overcome this difficulty is by using real monetary value as a common unit of input measure. Two means used for computing total-factor productivity are through index numbers and production functions (Lu, Cline, and Quance, 1979). The latter is most useful for the theoretical approach in this section.

A production function, as explained in Chapter 2, describes the physical relationship between input of resources and output per unit of time under a given state of technology. As technological change takes place the production function coefficients also change. Because changes are not continuous, smooth, or necessarily neutral, it is difficult to specify the functional form. However, the form can be conceptualized by visualizing an input mix on the horizontal axis of a graph rather than the usual single input being varied with the rest held constant as is typically depicted in a production function.

The production function for livestock, shown in Figure 12.2, has a production function P_0 using input mix X_0 yielding output Y_0 by producing at point a. Another alternative is use of X_2 input to produce Y_2 output. The impact of technological change can be considered in its simplest form as a movement from point a to point b on a new production P_1, at some later time by changing the mix of inputs. As an example, output Y_1 could result from purchase and use of a home computer tied to an individual cow ration distribution system wherein the

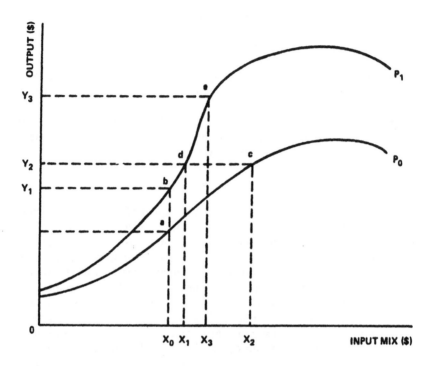

FIGURE 12.2   IMPACT OF TECHNOLOGICAL CHANGE ON OUTPUT.

discounted capital and maintenance costs are offset by a reduction in feed cost. An even higher output than Y_2 can be obtained by increasing the input mix. For example, if in addition to the home computer and feed distribution system, heat (estrus cycle) detectors were tied to the computer (X_3) an output of Y_3 at point e might be obtained. The important factor here is knowledge and careful manipulation of feedstuffs to optimize feed use.

A crucial point to understand future livestock industry technological change in developing as well as developed countries will be to recognize the new production functions will not be smooth, but rather will have "blips," as represented in Figure 12.2 by the segment d to e. In this case, a rather large output increase from Y_2 to Y_3 can take place at some point in the future with very little additional total input use, again because of expanded knowledge. Perhaps most important is to place technological change evaluation in its proper perspective by realizing that there are literally thousands of techniques and technologies available to livestock producers. Most technologies directly applicable to LDCs have been utilized in many parts of the world for a long time. But for all practical purposes these known technologies can be included with new technologies because they are new to a certain area and to those people. In effect, although much of the discussion on new technologies in this chapter has been about rather sophisticated technologies, the important point is awarness of the ongoing revolution in development of techniques and products, some of which are directly applicable to LDCs.

A vital first step is to determine what the potential techniques are, prioritize them in at least an approximate fashion (a first stage of ex-ante analysis), then determine the appropriate tool for the evaluation process and finally carry out an ex-ante economic analysis. Many of these analyses can be successfully carried out with relative little time expended, even though many assumptions are required provided the objective only is to prioritize potential technologies for future research. This process is analogous to the "sondeo" or rapid rural appraisal survey applied in the FSR/E methodology.

Technology advances and techniques to apply them that seem difficult or cumbersome today will be readily accessible in the not-too-distant future. Adoption will depend on effective demand for them--that is, on producers who have both economic stimulation and political encouragement. As an example of a simple ex-ante analysis let us now examine embryo transfer, a technology that will have widely variable adoption rates in the next decade. Evaluation of this technique not only provides a good illustration of how partial budgeting can be used for ex-ante analysis, but is also a prime illustration of the increasing importance that knowledge will play in animal agriculture.

Embryo Transfer: Example of Adoption Criteria

Embryo transfer (ET) is a technique that has only been adopted commercially since about 1980. The technologies are evolving so rapidly that major scientific papers have recently dealt with apparently mundane topics, such as advances in labeling procedures of straws holding frozen embryos. ET has been justified almost exclusively on the basis of genetic improvement, but great advances have been made in embryo transfer as a means to obtain multiple births. Let us discuss that use.

There are a number of procedures available to obtain multiple births, such as splitting embryos and transferring embryos to females already pregnant. Success rates are still widely variable, and thus estimates of the potential impact must be based on numerous assumptions, with a main one being national-level adoption rates. As an example, calculations in Table 12.1 show that if 20 percent of a nation's mature cattle females received transfers and had the potential for twins, a 70 percent success rate obtained, and those females not pregnant serviced the second time by the usual procedure of artificial insemination, the national calf crop born would increase 21 percent. If 60 percent of the females received transfers with the potential for twins, then calf crop would increase 63 percent. In effect, at the national level, there would be 97 calves born for every 100 cows if 20 percent of the females were involved in transfer rather than the present 80 calves. There would be 130 calves under the 60 percent transfer level. ET is indeed a potentially powerful technology available if there is adequate economic stimulation and proper political climate.

Japan is an example of a country where embryo transfer for multiple calving will likely be extensively adopted in the next few years. The country is very densely populated, and due to economic development Japan's self-sufficiency in beef production has dropped from 96 percent in the 1960s to 72 percent currently. These factors, plus some others exogenous to the cattle industry, have led to political determination that the domestic production level not drop much further.

The second overall factor--economics--also favors adoption of embryo transfer for multiple calving in Japan. Weaned calf prices are among the world's highest, about four times the calf price in the United States. As an example, using a beef breed calf production cost of $1,200 (U.S. dollars) as a base figure, and additional transfer-related cost of $400, total cost would be $1,600. But if the calf crop were to increase 63 percent (the high level shown in Table 12.1), then cost per calf would decrease by 13 percent to $1,046.[1] Transfer cost in Japan could easily fall to $200, which would result in a cost reduction of 26 percent. These cost reductions are sufficiently large that the techniques will likely be adopted by many producers, providing there is government interest.

National-level planners in Japan are, of course, also taking into account the more rapid genetic improvement possible in the national-level herd in addition to the potential for overcoming a shortfall in beef breed

Table 12.1 Potential national-level impact of embryo
transfers resulting in twins

Item	Percent of national herd females transferred with twins	
	20	60
Initial number of cows (head)	100	100
Twin Transfer Process		
Cows serviced by ET (cows)	20	60
Twin transfer success rate (percent)	70	70
Cows with twins (head)	14	42
Calves from ET (head)	28	84
Usual Breeding Method (remainder of cows)		
Usual procedure only (cows)	80	40
Residual of ET cows not pregnant (head)	6	18
Total to be bred (cows)	86	58
Usual method success rate (percent)	80	80
Calves from usual breeding (head)	69	46
Totals		
Total calves born (head)	97	130
Usual born calf crop (head)	80	80
Additional calves, both ET & usual calving (head)	17	50
Increased calf crop (percent)	21	63

Source: Simpson, Yoshida, Miyazaki and Kada (1985).

calves. Japan has had a substantial embryo transfer research program, with about 1 in every 600 cows being an ET recipient in 1984. In contrast, about 1 cow in every 980 in the United States was an ET recipient that year. The conclusion is that ET will become as common in Japan as artificial insemination within the next decade because there are both economic and political incentives.

An example from a developed country was deliberately chosen to stress that certain conditions are required for technology adoption. The assumptions, such as a 70 percent ET rate, are specific to a certain country and at a certain point in time. A conclusion is embryo transfer, despite much attention by a wide variety of technicians and planners, will have little applicability to most developing countries in the foreseeable future because basic economics are an unalterable constraint. Economics is the criterion that holds true for all technologies.

Implications of Technological Advances for
World Animal Agriculture

An idea of the potential impact the technological change will have on beef production in some countries can be obtained by review of the executive summary from an RCA symposium titled "Future Agricultural Technology and Resource Conservation" (English, Maetzold, Holding and Heady, 1983). The experts concluded that meat production per breeding cow and per breeding sow will increase 25 percent in the United States by the year 2000 and 60 percent by 2030. Broilers, already the target of vast productivity improvements, are expected to attain a 30 percent improvement in production efficiency by the year 2000. Age-to-market weight efficiency of catfish is expected to increase 20 percent by the year 2000 and 200 percent by 2030. New research, just in the past few years, indicates the 200 percent increase will be possible by 2000 (30 years sooner).

The conclusions, although directed at the United States, are relevant for even the poorest countries or regions of the world as the conclusions provide targets or possibilities from which planning goals can be set. The conclusions also demonstrate that the human factor (management) rather than technology is the basic constraint to providing most areas of the world with adequate animal products in the foreseeable future.

The brief summary of technologies either in the offing or that are in their early stages of adoption, plus the embryo transfer example, lead to the conclusion that the world is indeed entering an era in which knowledge will be of paramount importance. The human dimension is replacing capital as the critical factor. Rewards to both individuals and nations will come from imagination, proper planning, acceptable input/output price relationships, and a favorable political climate.

It appears that the greatest advances at the national level in animal productivity and efficiency enhancement can be derived from determination of techniques and technologies that offer the greatest potential to meet national-level goals. Once those techniques are identified, a program must be developed that ensures that all constraints to its adoption are overcome and that adequate support is provided to obtain the greatest synergistic effect. This concept might be called the technology-directed planning approach. It seems clear that human capital development will provide the greatest synergism regardless of the technology chosen.

It can be predicted with great confidence that burdensome supplies of animal products will be the rule in developed areas in the foreseeable future as a result of relatively low population growth, changes in food tastes and preferences, and new technology. At the opposite extreme, hunger and famine will prevail in many regions due to uncontrolled population growth, political fragility, periodic adverse climatic conditions and failure to invest in human capital. The problem is thus not one of inability at the global level to produce sufficient animal products to adequately feed a much larger world population than exists today. Rather, the difficulty is primarily lack of purchasing power, management capability and inappropriate policies at both the individual person and national levels in certain regions.

Producers are and will be the key to realization of improvements in animal agriculture. Management--and that means knowledge--is the weak link to improving the animal industry everywhere in the world. Technologies abound; the problem is to devise the means to accelerate their adoption. The rate in economically developed countries is already unbelievably fast due to entrepreneurship being properly utilized. The problem in low-income countries where adoption rates are less than satisfactory is to stimulate individual initiative. The key there, again, is to focus on the preliminary step, which is to determine client, (government or private sector), problems and work to provide compatible solutions.

Let us end by relating a curious event that happened in Japan during the Meiji era, as it will help to understand technological change during the next quarter century. It seems that during 1876-1877 an American, Dr. William S. Clark, president of Massachusetts Agricultural College, spent eight-and-a-half months in Hokkaido, now a major livestock production area, as president of Japan's first agricultural college. Although reference to him can scarcely be found in the United States, Dr. Clark's name and story are familiar to even the youngest Japanese student. This is partly because Clark was so successful in his pioneering work--but more important were his parting words. Dr. Clark's last day was spent in the country with his twenty-four students. After lunch and a walk around the area he mounted his horse, turned to them and shouted, "Boys, be ambitious!" He rode off, never looking back. My parting words a century later--at a time when international trade conflicts and societal change demand deep understanding of the world's agriculture is: "Boys, be imaginative!"

NOTES

1. Calculations of ET cost decrease based on 100 cows:

    1. 100 cows @ $1,200 per calf times 80 calves = $96,000 as a base cost
    2. ET cost of $400 per cow times 100 cows = $40,000 + $96,000 as base cost = $136,000
    3. New cost with ET ($136,000) divided by 130 calves (63 percent increase on base of 80 calves) = $1,046
    4. Percent decrease to new cost is $154 ÷ 1,200 = 13 percent

REFERENCES

Abassa, Kodjo Pierre. "Systems Approach to Gobra Zebu Production in Dohra, Senegal." Ph.D. Thesis, University of Florida, 1984.

Alderman, Rom, James R. Simpson, and Will Carlton. Beefloan: A Micro-computer Ranch Financial Analysis. Circular 624. Florida Cooperative Extension Service, University of Florida, September 1984.

Anderson, J.R. "Economic Models and Agricultural Production Systems." Proc. Aust. Soc. Anim. Prod. 9 (1972): 77-83.

Angirasa, Atiti Kumar. "Firm Level Beef Supply: A Simulation and Linear Programming Application in East Texas." Ph.D. Thesis, Texas A & M University, 1979.

Backus, W.R. "Greater Profits Possible with Companion Grazing of Sheep and Cattle." The Stockman 41 (May 1984):1, 22.

Baker, Frank. Proceedings. International Seminar on Ruminant Animal Production. Morrilton, Arkansas: Winrock International, January 10-11, 1983.

Bayley, N.D. "Research Resource Allocation in the Department of Agriculture." Resource Allocation in Agricultural Research, ed. W.L. Fisher, pp. 218-234. Minneapolis: University of Minnesota Press, 1971.

Beck, L. "Herd Owners and Hired Shepherds: The Qashqai of Iran." Ethnology 19 (1980): 327-352.

Beebe, James. Rapid Rural Appraisal: The Critical First Step in a Farming Systems Approach to Research. IFAS International Programs Farming Systems Support Project Networking Paper no. 5, University of Florida, 1985.

Berleant and R. Schiller. "The Social and Economic Role of Cattle in Barbudu." Geog. Rev 67 (1977): 200-309.

Blackburn, H.D., et.al. The Texas A & M Sheep and Goat Simulation Models. Texas Agri. Exp. Sta. Bull. no. B-1559, Texas A & M University, 1987.

Blackburn, H.D., T.C. Cartwright, P.J. Howard and F. Ruvuna. Development of Dual Purpose Goat Production for Smallholders in Western Kenya Using Computer Simulation and Systems Analysis. CSRS Technical Report Series no. 89. Texas Agr. Exp. Sta. Texas A & M University, December 1986.

Bonnemaire, J., and J.P. Deffontaines. "Factors Determining the Choice of Beef Production Systems: Implications for Land Use." The Future of Beef Production in the European Community, ed. J.C. Bowman and P. Susmel, pp. 51-73. Boston: Martinus Nijhoff, 1979.

Bonsembiante, M., and P. Susmel. "Feedlot Development in Italy." The Future of Beef Production in the European Community, ed. J.C. Bowman and P. Susmel, pp. 317-331. Boston: Martinus Nijhoff, 1979.

Boyd, H., and M. Koger. "Dynamo Simulation of Beef Cattle System Profitability." J. Anim. Sci. Abstract 39 (1974): 141.

Boykin, Calvin C., Jr. Economic and Operational Characteristics of Cattle Ranches. Texas Agri. Exp. Sta. Bull. no. MP-866, Texas A & M University, 1968.

Boykin, Calvin C., Jr., and Nathan K. Forrest. Economic and Operational Characteristics of Livestock Ranches: Edwards Plateau and Central Basin of Texas. Texas Agri. Exp. Sta. Bull. no. MP-978, Texas A & M University, 1971.

Boykin, Calvin C., Henry C. Gilliam, and Ronald A. Gustafson. Structural Characteristics of Beef Cattle Raising in the United States. USDA AER Report 450. Washington, D.C., 1980.

Bravo, B.F. "Beef Production Systems: A Simulation Approach." Ph.D. Thesis, University of New England (Armidale, NSW, Australia), 1972.

Bredahl, Maury E., Andrew Burst, and Philip F. Warnken. Growth and Structure of the Mexican Cattle Industry. College of Agriculture International Series 7, Special Report 317, University of Missouri, 1985.

Brokken, Ray, Carl O'Connor, and Thomas Norblom. Costs of Reducing Grain Feeding of Beef Cattle. AER 459, U.S. Dept. of Agr., August 1980.

Brown, Maxwell L. Farm Budgets: From Farm Income Analysis to Agricultural Project Analysis. World Bank Staff Occasional Paper no. 29. Baltimore: Johns Hopkins University Press, 1979.

Buccola, Steven, Ernest Bentley, and Warren Jessee. "The Role of Market Price-Weight Relationships in Optimal Beef Cattle Backgrounding Programs." Sou. J. Agri. Econ. 12 (1) (July 1980): 65-72.

Butterworth, M.H. Letter to Editor. World Animal Review 44 (1982): 44.

Byerlee, Derek, Larry Harrington, and Donald L. Winkleman. "Farming Systems Research: Issues in Research Strategy and Technology Design." Am. J. Agri. Econ. 65 (December) 1982): 897-904.

Camoens, J.K. "Asian Livestock Production and Management Systems." In Camoens, J.K. et al., Proceedings. Regional Workshop on Livestock Production Management, Manila July 9-17, 1984 Asian Development Bank. Manila, 1975.

Calo, L.L. et al. "Simultaneous Selection for Milk and Beef Production Among Holstein-Friesians." Journal of Dairy Science 56 (1973): 1080-1084.

Cartwright, T.C., F.G. Gomez, J.O. Sanders, and T.C. Nelson. "Simulated Milk-Beef Production Systems in Colombia." J. Anim. Sci. Abstract 45 (1977): 13.

Cartwright, T.C., and H.D. Blackburn. "Systems Analysis for Developing Goat Production." Proceedings. Fourth International Conference on Goats, Brasilia, March 8-31, 1987.

CATIE (Centro Agronomico Tropical de Investigacion y Ensenan za). Investigacion. Aplicada en Sistemas de Produccio. de Leche. Turrialba, Costa Rica, October 1983.

Chudleigh, P.D. "Problem Analysis, Hypothesis Testing and Project Identification: An Example." In J.A. Gartner (ed.). Proceedings FAO Expert Consultation on Improving the Efficiency of Small-Scale Livestock Production in Asia: A Systems Approach, held in Bangkok, December 6-10, 1983. Rome, 1984.

Clark, Nancy, Alvin Schupp, Thomas Bidner, and John Carpenter. Consumer Acceptance of Beef from Selected Breeds and Feeding Programs. D.A.E. Research Report no. 572, Dept. of Agr. Econ. and Agribus. Louisiana State University, Baton Rouge, November 1980.

Congleton, W.R., Jr., and R.E. Goodwill. "Simulated Comparisons of Breeding Plans for Beef Production--Part 1: A Dynamic Model to Evaluate the Effect of Mating Plan on Herd Age Structure and Productivity." Agr. Systems 5 (1980a): 207-219.

_____. "Simulated Comparisons of Breeding Plans for Beef Production--Part 2: Hereford, Angus and Charolais Sires Bred to Hereford, Angus and Hereford-Angus Dams to Produce Feeder Calves." Agri. Systems 5 (1980b): 221-232.

_____. "Simulated Comparisons of Breeding Plans for Beef Production--Part 3: Systems for Producing Feed Calves Involving Intensive Culling and Additional Breeds of Sire." Agri. Systems 5 (1980c): 309-318.

Conrad, Joseph H. "Confined vs. Nonconfined Animal Raising Within the Farming System." Proceedings 1983 Farming Systems Research Symposium, Kansas State University, Office of International Programs, May 1984.

Dahl, G., and H. Hjort. Having Herds--Pastoral Growth and Household Economy. Stockholm Studies in Social Anthropology, no. 2, University of Stockholm, 1976.

Davis, J.M., T.C. Cartwright, and J.O. Sanders. "Alternative Herd Management for Guyana." J. Anim. Sci. Abstract 43 (1976): 235-236.

de Alba, Jorge. "Progress in the Selection of the Latin American Dairy Criollo." World Animal Review 28 (1978): 26-30.

DeBoer, A.J., and D.E. Welsch. "Constraints on Cattle and Buffalo Production in a Northeastern Thai Village." Tradition and Dynamics in Small-Farm Agriculture, ed. Robert D. Stevens. Ames: Iowa State University Press, 1977.

DeBoer, A.J., M. Job, and G. Maundu. "The Relative Profitability of Meat Goats, Angora Goats, Sheep and Cattle in Four Agro-Economic Zones of Kenya." Paper presented at the Third International Conference on Goat Production and Disease, Tucson, Arizona, January 10-15, 1982.

DeBoer, A.J. (ed.) Ruminant Livestock in Intensive Agricultural Areas of Sichuan Province, China: Current Status and Development Projects. Morrilton, Arkansas: Winrock International, 1984.

DeBoer, John. UNDP/FAO Sheep and Goat Development Project, Kenya: Production Economics. AG:DP/KEN/75/002, Nairobi, 1981.

DeBoer, John. "Goat and Goat Product Market and Market Prospects: An International Perspective." Proceedings Third International Conference on Goat Production and Disease, University of Arizona, Tucson, January 10-15, 1982, pp. 37-40.

Delgado, Christopher. Livestock and Meat Marketing in West Africa (Mali). Center for Research on Economic Development (CRED), University of Michigan, 1980.

Delgado, Christopher, and John McIntire. "Constraints on Oxen Cultivation in the Sahel." Am. J. Agr. Econ. 64 (1982): 188-196.

Devendra, C. "Biological Efficiency of Milk Production in Dairy Goats." World Rev. Animal Production 11 (1975): 46.

Devendra, C. "The Role of Goats in Food Production Systems in Industrialized and Developing Countries." Proceedings. Fourth International Conference on Goats, Brasilia, March 8-13, 1987a.

Devendra, C. "Flock Management in Integrated Village Systems." Proceedings. Fourth International Conference on Goats, Brasilia, March 8-13, 1987b.

Devendra, C. "Feed Resources and Their Relevance in Feeding Systems for Goats in Developing Countries." Proceedings. Fourth International Conference on Goats, Brasilia, March 8-13, 1987.

Dillon, John L. "The Economics of Systems Research." Agricultural Systems 1 (1976): 5-22.

Doll, John P., and Frank Orazem. Production Economics: Theory with Applications. Columbus, Ohio: Grid, Inc., 1978.

Doyle, P.T., C. Devendra, and G.R. Pearce. Rice Straw as a Feed for Ruminants. Canberra: International Development Program of Australian Universitites and Colleges Limited, 1986.

Duckman, A.N., and G. B. Mansfield. Farming Systems of the World. London: Chatto and Windus, 1972.

Dyer, I.A., and C.C. O'Mary. The Feedlot. Philadelphia: Lea and Febiger, 1972.

Easterling, Edward, Dilso Negrette, and J. Richard Connor. Analysis of Alternative Beef Cattle Production-Marketing Systems for South Mississippi. Department of Agri. Econ. Res. Report, AEC no. 105. Mississippi State University, Starkville, October 1980.

English, Burton C., James A. Maetzold, Brian R. Holding, and Earl O. Heady. RCA Symposium: Future Agricultural Technology and Resources Conservation, Executive Summary. Ames, Iowa; Center for Agricultural and Rural Development, Iowa State University, 1983.

Erickson, Donald B., and Philip A., Phar. Guidelines for Developing Commercial Feedlots in Kansas. Manhattan: Kansas State University, Cooperative Extension Service, 1970.

FAO. Production Yearbook. Rome, 1986.

Fick, Gary W. "The Mechanism of Alfalfa Regrowth: A Computer Simulation Approach." Search: Agriculture 7 (3) 1977: 1-28.

Fick, Gary W. "A Pasture Production Model for Use in a Whole Farm Simulator." Agr. Systems 5 (1980): 137-161.

Fitzhugh, H.A. "Small Ruminants as Food Producers." New Protein Foods, vol 4. New York: Academic Press, 1981.

Fitzhugh, H.A. "Application of Systems Science to Beef Production. Proceedings. International Symposium on Beef Production, Kyoto, Japan, 1983.

Fitzhugh, F.A. "Application of Systems Science to Optimization of Cattle Breeding Schemes." EEC Seminar on Optimization of Cattle Breeding Schemes, Dublin, Ireland, November 1975.

Fitzhugh, F.A., and E.K. Byington. "Systems Approach to Animal Agriculture." World Animal Review 27 (1978): 2-6.

Fitzhugh, H.A., and J. DeBoer. Physical and Economic Constraints to Intensive Animal Production in Developing Countries. Occasional Publication no. 4, Slough, England: Society of Animal Production, 1981.

Fitzhugh, H. et al. Research on Crop-Animal Systems: Proceeding of a Workshop. Marrilton, Arkansas: Winrock International, June 1982.

Flora, Cornelia B., "Farming Systems Research and the Land-Grant System: Transferring Assumptions Overseas." FSR Background Paper no. 3. Farming Systems Research Paper Series. Office of International Agricultural Programs. Kansas State University, Oct. 1982.

Forrester, J.W. Principles of Systems. Cambridge, Mass.: Wright-Allen Press, 1973.

Frisch, J.E., and J.E. Vercoe. "Utilizing Breed Differences in Growth of Cattle in the Tropics." World Animal Review 25 (1978): 81-82.

Gall, C. ed. Goat Production, Orlando, Fla.: Academic Press, 1981.

Garner, J.K. Increasing the Work Efficiency of the Water Buffalo Through Use of Improved Harness. Washington, D.C.: USAID, Office of Agriculture, Development Support Bureau, Reprint, 1980.

Gartner, J.A. (ed.). Proceedings. FAO Expert Consultation on Improving the Efficiency of Small-Scale Livestock Production in Asia: A Systems Approach held in Bangkok, Thailand, December 6-10, 1983. Rome: FAO, 1984.

Gittinger, J. Price. Economic Analysis of Agricultural Projects. Baltimore: Johns Hopkins University Press, 1972.

Gladwin, Christina H. "Decision-Making of Small Producers on Hillside Farms and the Implications for Project Design." Paper presented at the Seminario Internacional Sobre Produccion Agropecuaria y Forestal en Zonas de Ladera de America Tropical, Turrialba, Costa Rica, December 1-5, 1980.

Goe, M.R. "Current Status of Research on Animal Traction." World Animal Review 45 (1983): 2-17.

Goodsell, Wylie D. Southwest Cattle Ranches: Organization, Costs and Returns, 1964-72. AER Bull. no. 255, US. Dept. of Agr., 1974.

Government of Australia, Bureau of Agricultural Economics. A Computer Model Simulating Extensive Beef Cattle Production Systems. Canberra: Occ. Paper no. 21, 1974.

Gray, James. Ranch Economics. Ames: Iowa State University Press, 1968.

Gray, Kenneth R. Soviet Livestock Cycles with United States Comparisons. Unpublished Report for National Council for Soviet and East European Research, January 1981.

Grigg, D.B. The Agricultural Systems of the World: An Evolutionary Approach. London: Cambridge University Press, 1974.

Groenwold, H.H., and P.R. Crossing. "The Place of Livestock in Small Farm Development." World An. Rev. 15 (1975): 2-6.

Gryseels, Guido. "Livestock in Farming Systems Research for Smallholder Agriculture: Experiences of ILCA's Highlands Programme." Seminar Paper, ILCA, 1983.

Gutierrez, Nestor F., and John DeBoer. "Marketing and Price Formulation for Meat Goats, Hairsheep and their Products in Ceara State, Northeast Brazil." Proceedings Third International Conference on Goat Production and Disease, University of Arizona, Tucson, Arizona, January 1-15, 1982, pp. 50-54.

Haenlein, G.F.W. "Dairy Goat Management." Journal of Dairy Science 61 (1978): 1011-1022.

Hallam, David. Livestock Development Planning: A Quantitative Framework. Center for Agriculture Strategy CAS Paper 12, University of Reading, 1983.

Halter, A.N., M.L. Hayenga, and T.J. Manetsch. "Simulating a Developing Agricultural Economy: Methodology and Planning Capability." Am. J. Agr. Econ. 52 (1970): 272-284.

Hansen, H.H., and R.D. Child. Goat Diets on an Upland Hardwood Ecosystem. Morrilton, Arkansas: Winrock International Livestock Research and Training Center, 1980.

Harris, Dewey L., Terry S. Stewart, and Cecilio R. Arbaleda. Animal Breeding Programs: A Systematic Approach to Their Design. Washington, D.C.: USDA AES Advances in Agricultural Technology Report ATTNC-8, February 1984.

Harris, M. Cows, Pigs, Wars and Witches: The Riddles of Culture. New York: Vintage Books, 1974.

Harsh, Stephen B., Larry J. Connor, and Gerald D. Schwab. Managing the Farm Business. Englewood Cliffs, N.J.: Prentice-Hall, Inc., 1981.

Hart, Robert, and R.E. McDowell. Crop/Livestock Interactions as: 1. Crop Production Determinants, 2. Livestock Production Determinants Cornell International Agriculture Mimeo 107, 1985.

Harwood, R.R. Small Farm Development: Understanding and Improving Farming Systems in the Humid Tropics. Boulder, Colo: Westview Press, 1979.

Heady, Earl O., and Wilfred Candler. Linear Programming Methods. Ames: Iowa State University Press, 1958.

Heady, Earl O., and Shashanka Bhide. Livestock Response Functions. Ames: Iowa State University Press, 1983.

Hildebrand, Peter E. "Modified Stability Analysis of Farmer Managed, On-Farm Trials." Agronomy Journal 76 (1984): 271-274.

Hildebrand, Peter E. "Motivating Small Farmers, Scientists and Technicians to Accept Change." Agricultural Administration 8 (1981a): 375-383.

Hildebrand, Peter E., and Federico Poey. On-Farm Agronomic Trials in Farming Systems Research and Extension. Boulder, Colo: Lynne Rienner Publishers, Inc., 1985.

Hildebrand, Peter E. "Combining Disciplines in Rapid Appraisal: The Sondeo Approach." Agricultural Administration 8 (1981b): 423-432.

Hildebrand, Peter E. "Farming Systems Research: Issues in Strategy and Technology Design: Discussion." Am. J. Agr. Econ. 65 (December 1982): 905-906.

Holdcraft, L.E. "The Rise and Fall of Community Development in Developing Countries, 1960-65." MSU Rural Development Paper, no. 2. East Lansing: Department of Agricultural Economics, Michigan State University, 1978.

Hoveland, C.S., et al. Forage Systems for Beef Cows and Calves in the Piedmont. Alabama Agr. Exp. Sta. Bull. no. 497, Auburn University, 1977.

Hunt, H.W. "A Simulation model for Decomposition in Grassland." Ecology 58 (1977): 469-484.

Huss, D.L. "Small Animals for Small Farms in Latin America." World Animal Review 43 (1982): 24-29.

Hussain, Tarig. The Use of SImulation in Appraising a Livestock Breeding/Fattening Project. Washington, D.C.: Int. Bank for Rec. and Dev. Rpt. no. 56, 1970.

ILCA. "Programs and Budget." Addis Ababa, Ethiopia, 1983.

Innis, George S., and Richard Miskimins. Ranges 1 Grassland Simulations Model. Regional Analysis of Grassland Environmental Systems Rpt. no. 8, Colorado State University, August 1973.

Inns, F.M. "Animal Power in Agricultural Production Systems." World Animal Review 34 (1980): 2-10.

International Bank for Reconstruction and Development, Computing Activities Department. A User's Guide to BLIVOO Herd Development Simulation with Cash Flow and Rate of Return. Washington, D.C., 1972.

(ILCA) International Centre for Livestock in Africa. Systems Study 3: Small Ruminant Production in Humid Tropics. Addis Ababa, Ethiopia, 1979.

(ILCA) International Centre for Livestock in Africa. Economic Trends: Small Ruminants. Bulletin 7. Addis Ababa, Ethiopia, 1980.

ILCA (International Centre for Africa). The Design and Implementation of Pastoral Development Projects for Tropical Africa. Working Document Y. Addis Ababa, Ethiopia, 1980.

(ILCA) International Livestock Center for Africa. Mathematical Modeling of Livestock Production Systems: Application of Texas A & M University Beef Cattle Production Model for Botswana. ILCA System Study no. 1, Addis Ababa, Ethiopia, 1978.

Ive, J.R. "Simulating Change in Cattle Liveweight During the Transition Period in a Dry Monsoonal Climate." Aust. J. Agr. Res. 27 (1976): 297-307.

Jahanke, Hans E. Livestock Production Systems and Livestock Development in Tropical Africa. Kiel, West Germany: Kieler Wissenschaft-Sverlag Vauk, 1982.

Johnson, S.R., and G.C. Rausser, "Systems Analysis and Simulation: A Survey of Applications in Agricultural and Resource Economics." A Survey of Agricultural Economics Literature, vol. 2, ed. Lee R. Martin, pp. 157-301. Minneapolis: University of Minnesota Press, 1977.

Josserond, Henri P., and Edgar J. Ariza-Nino. "The Marketing of Small Ruminants in West Africa." Proceedings Third International Conference on Goat Production and Disease, University of Arizona, Tucson, Arizona, January 10-15, 1982, pp. 55-62.

Juri, Patricia, Nestor F. Gutierrez, and Alberto Valdes. Modelo de Simulacion por Computador para Fincas Ganaderas. Cali, Colombia: Centro Internacional de Agricultura Tropical, August 1977.

Kapture, Judy. "An Overview of Problems in Marketing Dairy Goat Products in the U.S.A." Proceedings. Third International Conference on Goat Production and Disease, University of Arizona, Tucson, Arizona, January 10-15, 1982, pp. 63-67.

Kay, Ronald D. Farm Management: Planning Control, and Implementation. New York: McGraw-Hill, Inc., 1981.

Kemeny, John G. A Philosopher Looks at Science. New York: D. Van Nostrand, Co., Inc., 1959.

Keulen, H. van. "Simulation of Water Use and Herbage Growth in Arid Regions." Simulation Monographs. Pudoc, Wageningen, 1975.

Khan, H.D. "Ten Decades of Rural Development: Lesson from India." MSU Rural Development Paper, no. 1. East Lansing: Department of Agricultural Economics, Michigan State University, 1978.

Kiflewahid, Berhane, Gordon R. Potts, and Robert M. Drysdale, eds. By-Product Utilization for Animal Production: Proceedings of a Workshop on Applied Research Held in Nairobi, Kenya, 26-30 September 1982. Ottawa: International Development Research Centre, 1983.

Knipscheer, Hendrick, Robert D. Hart, and Greg Baker. "Socioeconomic Aspects of Small Ruminant Activity." Proceedings. Fourth International Conference on Goats, Brasilia, March 8-13, 1987.

Koch, R.M., and J.W. Algeo. "The Beef Cattle Industry: Changes and Challenges." J. An. Sci. 57, Supp. 2 (July 1983): 28-43.

Kolors, J.F., and D. Bell. Physical Geography--Environment and Man. New York: McGraw-Hill, 1975.

Konandreas, Panos A., and Frank M. Anderson. Cattle Herd Dynamics: An Integer and Stochastic Model for Evaluating Production Alternatives. Unpublished report. Addis Ababa, Ethiopia: International Livestock Centre for Africa, September 1980.

Kottke, Marvin W. "Budgeting and Linear Programming Can Give Identical Solutions." J. Farm Mgt. 43 (1961): 307-314.

Levine, Joel M., William Hohenboken, and A. Gene Nelson. "Simulation of Beef Cattle Production Systems in the Llanos of Colombia--Part I. Methodology: An Alternative Technology for the Tropics." Agr. Systems 7 (1981): 37-48.

Levine, Joel M., and William Hohenboken. "Simulation of Beef Cattle Production Systems in the Llanos of Columbia--Part II. Results of the Modeling." Agr. Systems 7 (1981): 83-93.

Little, Peter D. "Critical Socio-Economic Variables in African Pastoral Livestock Development: Toward a Comparative Framework." Livestock Development in Subsaharan Africa: Constraints, Prospects, Policy, ed. James R. Simpson and Phylo Evangelou, pp. 201-214. Boulder, Colo.: Westview Press, 1984.

Long, C.R., T.C. Cartwright, and H.A. Fitzhugh, Jr. "Systems Analysis of Sources of Genetic and Environmental Variation in Efficiency of Beef Production: Cow Size and Herd Management." J. Anim. Sci. 40 (1975): 409-420.

Lu, Yao-Chi, Philip Cline, and LeRoy Quance. Prospects for Productivity Growth in U.S. Agriculture. Washington, D.C.: USDA ESCS Agr. Econ. Rpt. 4 35, September 1979.

McCown, R.L. "The Climatic Potential for Beef Cattle Production in Tropical Australia: Part I--Simulating the Annual Cycle of Liveweight Change." Agr. Systems 6 (1980): 303-317.

McCown, R.L., P. Gillard, L. Winks, and W.T. Williams. "The Climatic Potential for Beef Cattle Production in Tropical Australia: Part II-- Liveweight Change in Relation to Agroclimatic Variables." Agr. Systems 7 (1981a): 1-10.

McCown, R.L., P. Gillard, L. Winks, and W.T. Williams. "The Climatic Potential for Beef Cattle Production in Tropical Australia: Part III--Variation in the Commencement, Cessation and Duration of the Green Season." Agr. Systems 7 (1981b): 163-178.

McDowell, R.E. "Role of Animals in Developing Countries." Animals. Feed. Food and People. AAAS Symposium Volume, Washington, D.C., 1979.

McDowell, R.E., and P.E. Hildebrand. Integrated Crop and Animal Production: Making the Most of Resources Available to Small Farms in Developing Countries. New York: The Rockefeller Foundation Working Papers, January 1980.

McDowell, R.E., and L. Bove. The Goat as a Producer of Meat. Cornell Inter. Agr. Mimeo no. 56, Cornell University, 1977.

McGrath, J.E., Nordlie, P.G., and W.S. Vaughn. "A Descriptive Framework for Comparison of System Research Methods." Systems Analysis, ed. W.L. Fisher, pp. 73-86. New York: Penguin, 1973.

McKissick, John C., and R. Edward Brown, Jr. Georgia-Stocker-Finishing Alternatives: Fall/Winter 1981-82. Special Report 205. Management and Marketing Department, Cooperative Extension Service, University of Georgia, 1981.

Merrill, L.B. "The Role of Goats in Biological Control Brush." Rangeland Resources Research. Consolidated Progress Report 3665, Texas Agr. Exp. Sta., 1980, p. 21.

Miller, W.C., J.S. Vrinks, and T.M. Sutherland. "Computer Assisted Management Decisions for Beef Production Systems." Agr. Systems 3 (1978): 147-158.

Minish, Gary I., and Danny G. Fox. Beef Cattle Production and Management, 2nd ed. Reston, Va.: Reston Publishing Company, 1982.

Moore, Patrick. "Early Weaning for Increased Reproduction Rates in Tropical Beef Cattle." World Animal Review 49 (1984): 39-50.

M'Pia, Elengasa, and James R. Simpson. "Use of Economic-Demographic Simulation Model for Planning in Zaire's Beef Cattle Subsector. Journal of African Studies 11 (Summer 1984): 75-83.

284

Myen, Klaus. "Ranching in Kenya: Fattening of Steers or Cow-Calf Operation." Quarterly Journal of International Agriculture 17 (October-December 1979): 353-359.

National Research Council. The Water Buffalo: New Prospects for an Under-utilized Animal. Washington, D.C.: National Academy Press, 1981.

Nelson, T.C., T.C. Cartwright, and J.O. Sanders. "Simulated Production Efficiencies from Biologically Different Cattle in Different Environments." J. Anim. Sci. 47 (Supplement 1) (1978): 60.

Nordblom, Thomas L., Awad El Karim, Hamid Ahmed, and Gordon R. Potts. Research Methodology for Livestock On-Farm Trials. Ottawa: International Development Research Centre, 1985.

Norman, David, and Elan Gilber. "A General Overview of Farming Systems Research." Readings in Farming Systems Research and Development ed. W.W. Shaner, P.F. Philipp, and W.R. Schmeh, pp. 18-34. Washington, D.C.: Consortium for International Development, 1981.

Norman, D.W., Pryor, D.H., and Gibbs, C.J.N. "Technical Change and the Small Farmer in Housaland, Northern Nigeria." African Rural Economy Paper, no. 21. East Lansing: Department of Agricultural Economics, Michigan State University, 1979.

Norman, D.W. "The Farming Systems Approach: Relevancy for the Small Farmer." MSU Rural Development Paper, no. 5 East Lansing: Department of Agricultural Economics, Michigan State University, 1980.

Norman, David W. "The Farming Systems Approach to Research." FSR Background Paper no. 3. Farming Systems Research Paper Series. Office of International Agricultural Programs, Kansas State University, October 1982.

Notter, D.R. "Simulated Efficiency of Beef Production for a Cow-Calf Feedlot Management System." Ph.D. Thesis, University of Nebraska, 1977.

Oltenacu, E., A. Martinez, H. Glimp, and H. Fitzhugh, eds. The Role of Sheep and Goats in Agricultural Development. Proc. Workshop. Morrilton, Arkansas: Winrock Inter. Livestock Res. and Training Ctr., 1976.

Ordonez, J. "Systems Analysis of Beef Production in the Western High Plains of Venezuela." Ph.D. Thesis, University of Nebraska, 1977.

Osburn, Donald D., and Kenneth C. Schneeberger. Modern Agricultural Management: A Systems Approach to Farming, 2nd ed. Reston, Va.: Reston Publishing Co., Inc., 1983.

Pate, F.M., and J.R. Crockett. Value of Preconditioning Beef Calves. Florida Agri. Exp. Sta. Bull. no. 799, University of Florida, 1979.

Peake, D.C.I., E.F. Henzell, and G.B. Stirk. "Simulation of Changes in Herbage Biomass and Drought Response of a Buffel Grass (Cenchrus cilarus cv. Biloela) in Southern Queensland". Agroecosystems 5 (1979): 23-40.

Pearson de Vaccaro, Lucia, ed. Sistemas de Produccion con Bovino en el Tropico Americano. Proceedings of a workshop in Venezuela, June 1981. Facultad de Agronomia, Universidad Central de Venezuela, Maracay, Venezuela, 1982.

Penson, Jr., John B., Danny A. Klinefelter, and David A. Lins. Farm Investment and Financial Analysis. Englewood Cliffs, N.J.: Prentice-Hall, Inc., 1982.

Perry, Tilden Wayne. Beef Cattle Feeding and Nutrition. New York: Academic Press, 1980.

Petritz, David C., Steven P. Erickson, and Jack H. Armstrong. The Cattle and Beef Industry in the United States: Buying, Selling, Pricing. CES Paper 93. Cooperative Extension Service, Purdue University, 1982.

Poitevin, J., M. Mallard, and E. Picon. Beef Production in Southern Europe. Paris: OECD, 1977.

Poleman, T.T., and Freebairn, U.K., eds. Food, Population and Employment. New York: Praeger, 1973.

Potts, Gordon. "Application of Research Results on By-Product Utilization: Economic Aspects to be considered." By-Product Utilization for Animal Production. Proceedings of a Workshop on Applied Research held in Nairobi, Kenya September 26-30, 1982, Ottawa, Canada, IDRC, 1983.

Powers, Terry A. HERDSIM Simulation Model: User Manual. Washington, D.C.: InterAmerican Development Bank Project Analysis Monograph no. 2, February 1975.

Preston, T.R., and M.B. Willis. Intensive Beef Production, 2nd ed. New York: Pergamon Press, 1974.

Production Credit Associations. Agriculture 2000: A Look at the Future. Columbus, Ohio: Battelle Press, 1983.

Puterbaugh, H.L. "An Application of PPB in the Agricultural Research Service." Resource Allocation in Agricultural Research, ed. W.L. Fisher, pp. 218-234. Minneapolis, University of Minnesota Press, 1971.

Ramaswamy, N.S. The Management of Animal Energy Resources and the Modernization of the Bullock-Cart System. Bangalore: Indian Institute of Management, 1979.

Raun, Ned S. "The Emerging Role of Goats in World Food Production." Proceedings. Third International Conference on Goat Production and Disease, University of Arizona, Tucson, Arizona, January 10-15, 1982, pp. 133-140.

Raun, N.S., and K.L. Turk. "International Animal Agriculture: History and Perspectives." J. of An. Sci. 57 Supp. 2 (July 1983): 156-170.

Reh, Ingeborg, and Peter Horst. "Beef Production from Draught Cows in Small-Scale Farming." Q.J. Inter. Agr. 24 (March 1985): 38-47

Reid, R.L., and T.J. Klopfenstein. "Forages and Crop Residues: Quality Evaluation and Systems of Utilization." J. An. Sci. 57 Supp. 2 (July 1983): 534-562.

Rendall, D.F., and B.A. Lockwood. "The Economics of Two Major Dairy Systems in the Punjab, Pakistan." In J.C. Fine and R.G. Lattimore (eds.). Livestock in Asia: Issues and Policies. Ottawa, Canada: IDRC 1982, pp. 54-69.

Richards, Jack A., and Karzan Gerald E. Beef Cattle Feedlots in Oregon: A Feasibility Study. Agri. Exp. Station Spec. Rep. 170. Corvallis, Oregon State University, 1974.

Richardson, Harry W. "The State of Regional Economics: A Survey Article." International Regional Science Review 3 (1978): 1-48.

Richardson, William W., William G. Camp, and William G. McVay. Managing the Farm and Ranch. Reston, Va.: Reston Publishing Co., Inc., 1982.

Ruthenberg, H. Farming Systems in the Tropics. Oxford: Clarendon Press, 1971.

Salmon, David G., and Phillip F. Warken. Economics of Milk Product in Mexico's Humid Tropics, Special Report 286, International Small Farm Program, International Series 5, University of Missouri, October 1982.

Sanders, J.O. "A Model of Reproductive Performance in the Bovine Female." M.S. Thesis, Texas A & M University, 1974.

Sanders, J.O., and T.C. Cartwright. "A General Cattle Production Systems Model. I. Description of the Model." Agr. Systems 4 (January 1979a): 217-227.

Sanders, J.O., and T.C. Cartwright. "A General Cattle Production Systems Model. II. Procedures Used for Simulating Animal Performance." Agr. Systems 4 (January 1979b): 289-309.

Sanders, J.O. "Application of a Beef Cattle Production Model to the Evaluation of Genetic Selection Criteria." Ph.D. Thesis, Texas A & M University, 1977.

Sandford, S. Pastoralism and Development in Iran. Pastoral Network Paper 3c, Overseas Development Institute, London, 1977.

Sandford, Stephen. "Institutional and Economic Issues in the Development of Goat Product Markets." Proceedings. Third International Conference on Goat Production and Disease, University of Arizona, Tucson, January 10-15, 1982, pp. 31-35.

Sands, Michael, and Robert E. McDowell. A World Bibliography on Goats. Cornell International Agriculture Mimeograph 70, Cornell University, 1979.

Sands, Michael, James Simpson, Luis Hertaintains, and Lee McDowell. "Evaluation of Mineral Supplements for Grazing Cattle on Cattle Farms in Chiriqui, Panama." Unpublished report, November 1985.

Schaefer-Kehner, Walter. "Economic Aspects of Intensive Beef Production in a Developing Country." Quarterly Journal of International Agriculture 17 (October-December,1978): 342-352.

Schank, Roger, and Robert Abelson. Scripts, Plans, Goals and Understanding. New York: Wiley and Sons, 1977.

Schillhorn-van-Veen, T.W. Notes and Observations on the Livestock Sector in the Margui-Wandala Area of North Cameroon. East Lansing, Mich.: MSU/USAID Mandara Mountain Research Report no. 4, June 1980.

Schneider, Harold K. "Livestock in African Culture and Society: A Historical Perspective." Livestock Development in Subsaharan Africa: Constraints, Prospects, Policy. ed. James R. Simpson and Phylo Evangelou, pp. 187-200. Boulder, Colo.: Westview Press, 1984.

Sere, C., and W. Doppler. "Simulation of Production Alternatives in Ranching Systems in Togo." Agr. Systems 6 (1980-81): 249-260.

Shaner, W.W., P.F. Philipp, and W.P. Schmehl. Farming Systems Research and Development: Guidelines for Developing Countries Boulder, Colo.: Westview Press, 1981.

Sheldon, B. "Goats Don't Stop Woody Weeds." Rural Research. Commonwealth Scientific Industrial Research Organization, 109 (1980): 7-10.

Sheton, M. "Reproduction and Breeding of Goats." J. Dairy Science 61 (1978): 994-1010.

Simpson, James R. "Costs and Returns in a Study of Common Property Range Improvements." J. Range Mgt. 24 (1971): 248-251.

Simpson, James R., L.B. Baldwin, and F.S. Baker, Jr. Investment and Operating Costs for Two Types and Three Sizes of Florida Feedlots. Bull. 817. Fla. Agr. Exp. Sta., January 1981.

Simpson, James R., and Forest E. Stegelin. "An Economic Analysis of the Effect of Increasing Transportation Costs on Florida's Cattle Feeding Industry." Sou. J. Agr. Econ. 13, no. 2 (1981): 141-148.

Simpson, James R., and F.S. Baker, Jr. "Methods for Carrying Out Economic Analyses for Beef Cattle Feeding Trials." Proceedings. Sixteenth Annual Conference on Livestock and Poultry in Latin America, May 9-14, 1982, B38-B54.

Simpson, James R., and Donald E. Farris. The World's Beef Business. Ames: Iowa State University Press, 1982.

Simpson, James R. "Identification of Goals and Strategies in Designing Technological Change for Developing Countries." Transferring Food Technology to Developing Nations: Economic and Social Perspectives, ed. Molnor, J.J., and H.A. Clonts, pp. 29-41. Boulder, Colo.: Westview Press, 1983.

Simpson, James R. "Problems and Constraints, Goals and Policy: Conflict Resolution in Development of Subsaharan Africa's Livestock Industry," pp. 5-20 in James R. Simpson and Phylo Evangelou (ed.). Livestock Development in Subsaharan Africa: Constraints, Prospects, Policy, Boulder, Colo.: Westview Press, 1984a.

Simpson, James R. Technological Changes That Will Affect the Cattle Industry. Staff Paper 263. Food and Resource Economics Department, University of Florida, September 1984b.

Simpson, James R. "Beef Prices and Grading Systems: A Worldwide Perspective." World Animal Review 50 (April-June) 1984c: 29-35.

Simpson, James R., and Aubry Bordelon. "The Production-Marketing Connection in Florida's Beef Industry: A Historical Perspective." The Florida Cattlemen 48 (5) (1984): 50-52.

Simpson, James R., and John DeBoer. "Project Design for Mixed Farming Systems in Northeast Thailand." Proceedings of Kansas State University's 1983 Farming Systems Research Symposium, Animals in the Farming System, ed. Cornelia Butler Flora, pp. 485-504. Paper no. 6, Office of Internation Programs, Kansas State University, May 1984.

Simpson, James R., and Phylo Evangelou (eds.). Livestock Development in Subsaharan Africa: Constraints, Prospects, Policies. Boulder, Colo.: Westview Press, 1984.

Simpson, James R., and Gregory M. Sullivan. "Planning for Institutional Change in Utilization of Subsaharan-Africa's Common Property Range Resources." African Studies Review 27 (4) (December 1984): 61-78.

Simpson, James R. "Projected Impacts of Technological Change on the EEC's Livestock Industry." Feedstuffs, January 14, 1985, pp. 22-25.

Simpson, James R., Tadashi Yoshida, Akira Miyazaki, and Ryohei Kada. Technological Changes in Japan's Beef Industry Boulder, Colo.: Westview Press, 1985.

Simpson, James R., William Kunkle, Robert Sand and Gene Cope. Beef Cattle Production Practices That Pay--South and Central Florida. Bull 198 Florida Cooperative Extension Service, 1984,

Simpson, James R., and Jimmye S. Hillman. "Influence of Technological Change of Future United States Beef Exports to the EEC." Q.J. Inter. Agr. 24 (1985): 279-292.

Simpson, James R., and Robert E. McDowell. "Livestock in the Economies of Sub-saharan Africa," in (ed.) Hansen, Art and Della E. McMillan, Food in Sub-Saharan Africa, pp. 207-221. Boulder, Colo.: Lynne Rienner Publishers, Inc., 1986.

Simpson, James R. "Economics of Ralgro." Unpublished paper presented at the Ralgro Conference in connection with the International Stockman's School, Houston, Texas, February 18, 1987.

Smith, G.M. "The TAMU Beef Cattle Production Model." An. Sci. Dept., Texas A & M University, 1979.

Smith, A.J. "Draught Animal Research: A Neglected Subject." World Animal Review 40 (1981): 43-48.

Smith, Blair J. The Dairy Cow Replacement Problem: An Application of Dynamic Programming. Bul. 745 (Technical). Fla. Agr. Exp. Sta., Univ. of Florida, April 1971.

Sollod, Albert E., Katherine Wolfgang, and James A. Knight. "Veterinary Anthropology: Interdisciplinary Methods in Pastoral Systems Research." Livestock Development in Subsaharan Africa: Constraints, Prospects, Policy, ed. James R. Simpson and Phylo Evangelou, pp. 285-302. Boulder, Colo.: Westview Press, 1984.

Sprague, H.B. "Combined Crop/Livestock Farming Systems for Developing Countries of the Tropics and Subtropics." Agricultural Technology for Developing Countries. Tech. Series Bull. no. 19. Washington, D.C.: TAB/USAID, 1976.

Spreen, T.H., and D.H. Laughlin. Simulation of Beef Cattle Production Systems and its Use in Economic Analysis. Boulder, Colo.: Westview Press, 1986.

Spring, Anita. "Men and Women Smallholder Participants in a Stall Feeder Livestock Program in Malawi." Kansas State University's 1983 Farming Systems Research Symposium. FSR paper no. 6, pp. 726-742.

Squire, H.A. "Experiences With the Development of an Intensive Beef Feedlot System in Kenya." In ed. A.J. Smith. Beef Cattle Production in Developing Countries. Edinburgh, Scotland: University of Edinburgh, Center for Tropical Veterinary Medicine, 1972, pp. 150-153.

Squire, H.A., and M.J. Creek. "Custom Feeding of Cattle--A Proposal for a Standardized Package Approach to Project Formulation." World Animal Review 8 (1973): 8-12.

Stoin, H.R. "Controlling Brush with Goats." Arkansas Farm Res. 19 no. 4 (1970): 12.

Stokes, Kenneth W., Donald E. Farris, and Thomas C. Cartwright. "Economics of Alternative Beef Cattle Genotype and Management/Marketing Systems." Sou. J. Agr. Econ. no. 2 (December 1981): 1-10.

Studemann, John A. (ed.). Forage-Fed Beef: Production and Marketing Alternatives in the South. Southern Cooperative Series Bull. 220, June 1977.

Sullivan, Gregory M. "Impact of Government Policies on the Performance of the Livestock-Meat Subsector." Livestock Development in Sub-saharan Africa: Constraints, Prospects, Policy, ed. James R. Simpson and Phylo Evangelou, pp. 143-160. Boulder, Colo.: Westview Press, 1984.

Sullivan, Gregory M., Donald E. Farris, and James R. Simpson. "Planning for Institutional Change in Utilization of Sub-saharan Africa's Common Property Range Resource." African Studies Review 27 (1984): 61-78.

Swift, T. "The Development of Livestock Trading in a Nomad Pastoral Economy: The Somali Case." Pastoral Production and Society, ed. L'Equire Ecologie et Anthropologie Des Socities Pastorales. Cambridge: Cambridge University Press, 1979.

Swift, T. "The Economics of Traditional Nomadic Pastoralism: The Tuareg of the Adrar N Iforas (Mali)." Ph.D. Thesis, University of Sussex (U.K.), 1981.

Teele, Thurston F. "Development and Management of Livestock Projects in the Sahel Area of Africa." Livestock Development in Sub-saharan Africa: Constraints, Prospects, Policy ed. James R. Simpson and Phylo Evangelou, pp. 225-240. Boulder, Colo.: Westview Press, 1984.

Tillman, Allan D. Animal Agriculture in Indonesia. Morrilton, Arkansas: Winrock International, 1981.

Tokrisna, Ruangrai, and Theodore Panayotou. "An Overview of the Livestock Sector in Thailand with Special Reference to Buffalo and Cattle." In J.C. Fine and R.G. Lattimore (eds). Livestock in Asia: Issues and Policies. Ottawa, Canada: IDRC, 1982: 130-137.

Turner, H.W. "Sheep and the Small Holder." World An. Rev. 28 (1978): 4-8.

U.S. Dept. of the Interior, Bureau of Land Management. Southern Rio Grande Planning Area, New Mexico: Environmental Impacts Statement. Washington, D.C.: May 2, 1981a.

U.S. Dept. of the Interior, Bureau of Land Management. Domestic Livestock Grazing Management Program for the Sonoma-Gerlach Resource Area, Nevada. Washington D.C.: March 25, 1981b.

Valdez, Alberto, and Gustavo Nores. Growth Potential of the Beef Sector in Latin-America--Survey of Issues and Policies. Washington, D.C.: International Food Policy Institute. 1979.

Von Soest, Peter J. "Impact of Feeding Behaviour and Digestive Capacity on Nutritional Response." Paper presented at the Technical Consultation on Animal Genetic Resources Conservation and Management, Rome Italy, 1980.

Walrath, Arthur J. Evaluation of Investment Opportunities: Tools for Decision Making in Farming and Other Businesses. ERS Agr. Handbook no. 349. U.S. Dept. of Agriculture, June 1977.

Ward, Clement E. Contract Integrated, Cooperative Cattle Marketing System. Mkt. Res. Rpt. no. 1078, U.S. Dept. of Agr., July 1977.

Wardle, C. "Smallholder Fattening of Beef Cattle in the Niger: Results from a Pilot Scheme." World Animal Review 32 (1979): 14-17.

Watts, Ronald. "Goats in Africa." Livestock International 10 (April/May 1982): 34-36.

White, J. The Estimation and Interpretation of Pastoralists' Price Responsiveness. Pastoral Network Paper 11f, Overseas Development Institute, London, 1981.

White, T.K. et al. "A Systems ANalysis of the Guyanese Livestock Industry." Agr. Systems 3 (1978): 47-66.

Whitson, Robert E., Don L. Parks, and Dennis B. Herd. "Effects on Forage Quality Restrictions on Optimal Production Systems Determined by Linear Programming." Sou. J. Agr. Econ. 8 (2) (1976): 1-4.

Williams, Ed, and Donald E. Farris. Economics of Beef Cattle Systems--From Weaning Age to Slaughter. Dept. Info. Rpt. no. 74-3. Department of Agricultural Economics and Rural Sociology, Texas A & M University, 1974.

Wilson, R.T. "The Integration of Goats in the Livestock Systems of Arid and Semi-Arid Africa." Proceedings. Fourth International Conference on Goats, Brasilia, March 8-13, 1987.

Wilton, J.W., C.A. Morris, E.A. Jenson, A.O. Leigh, and W.C. Pffifer. "A Linear Programming Model for Beef Cattle Production." Canadian J. Anim. Sci. 54 (1974): 693-699.

Winrock International. "Livestock Program Priorities and Strategy." Draft position paper prepared for the U.S. Agency for International Development. December 10, 1981.

Winrock International. Bibliography on Crop-Animal Systems. Morrilton, Arkansas, 1982a.

Winrock International. Case Studies for a Workshop: Research on Crop-Animal Systems. Morrilton, Arkansas, 1982b.

Wright, A. et al. "An Evaluation of the Role of Systems Modelling in an Agricultural Research Programme." Proc. N.Z. Anim. Prod. 56 (1976): 150-160.

Ying, Jiang, An Min, and Wang Shiquan. "Goat Production in China." Proceedings. Fourth International Conference on Goats, Brasilia, March 8-13, 1987.

INDEX